COLLEGE OF AEROSPACE DOCTRINE, RESEARCH AND EDUCATION

AIR UNIVERSITY

GPS versus Galileo
Balancing for Position in Space

Scott W. Beidleman
Lieutenant Colonel, USAF

CADRE Paper No. 23

Air University Press
Maxwell Air Force Base, Alabama 36112-6615

May 2006

Air University Library Cataloging Data

Beidleman, Scott W.
　　GPS versus Galileo : balancing for position in space / Scott W. Beidleman.
　　　　p. ; cm. – (CADRE paper, 1537-3371 ; 23)
　　Includes bibliographical references.

　　1. Global Positioning System. 2. Galileo Joint Undertaking. 3. Artificial satellites in navigation. 4. Astronautics and state. I. Title. II. Series. III. Air University (U.S.). College of Aerospace Doctrine, Research and Education.

623.893—dc22

Disclaimer

Opinions, conclusions, and recommendations expressed or implied within are solely those of the author and do not necessarily represent the views of Air University, the United States Air Force, the Department of Defense, or any other US government agency. Cleared for public release: distribution unlimited.

This CADRE Paper and others in the series are available electronically at the Air University Research Web site http://research.maxwell.af.mil and the AU Press Web site http://aupress.maxwell.af.mil.

CADRE Papers

CADRE Papers are occasional publications sponsored by the Airpower Research Institute of Air University's College of Aerospace Doctrine, Research and Education (CADRE). Dedicated to promoting the understanding of air and space power theory and application, these studies are published by Air University Press and broadly distributed to the US Air Force, the Department of Defense and other governmental organizations, leading scholars, selected institutions of higher learning, public-policy institutes, and the media.

All military members and civilian employees assigned to Air University are invited to contribute unclassified manuscripts that deal with air and/or space power history, theory, doctrine or strategy, or with joint or combined service matters bearing on the application of air and/or space power.

Authors should submit three copies of a double-spaced, typed manuscript and an electronic version of the manuscript on removable media along with a brief (200-word maximum) abstract. The electronic file should be compatible with Microsoft Windows and Microsoft Word—Air University Press uses Word as its standard word-processing program.

Please send inquiries or comments to
Chief of Research
Airpower Research Institute
CADRE
401 Chennault Circle
Maxwell AFB AL 36112-6428
Tel: (334) 953-5508
DSN 493-5508
Fax: (334) 953-6739
DSN 493-6739
E-mail: cadre.research@maxwell.af.mil

Contents

Chapter		Page
	DISCLAIMER	ii
	FOREWORD	vii
	ABOUT THE AUTHOR	ix
	ACKNOWLEDGMENTS	xi
1	INTRODUCTION	1
	Notes	9
2	GPS VERSUS GALILEO	11
	Notes	27
3	WHY GALILEO?	31
	Notes	45
4	IMPLICATIONS AND RECOMMENDATIONS	51
	Notes	66
	ABBREVIATIONS	69
	BIBLIOGRAPHY	71

Illustrations

Figure		
1	Satellite geometry	9
2	Global positioning system (GPS) satellite	14
3	Galileo satellite	16

Table

1	Number of visible satellites for various masking angles	13

Foreword

This study investigates Europe's motives to develop the independent satellite navigation system known as Galileo despite the existence of America's successful global positioning system (GPS). The study begins by analyzing both systems to familiarize the reader with global navigation satellite systems (GNSS) and to provide an understanding of the strengths and weaknesses of GPS and Galileo, as well as the systems' similarities and differences. Although the two systems have different founding principles, they employ similar infrastructures and operational concepts. In the short term, Galileo will provide better accuracy for civilian users until GPS upgrades take effect. But performance is only part of the rationale. The author contends that Europe's pursuit of Galileo is driven by a combination of reasons, including performance, independence, and economic incentive. With Galileo, Europe hopes to achieve political, security, and technological independence from the United States. Additionally, Europe envisions overcoming the US monopoly on GNSS by seizing a sizable share of the expanding GNSS market and setting a new world standard for satellite navigation. Finally, the author explores Galileo's impact on the United States and reviews US policy towards Galileo. The study concludes with recommendations to strengthen the competitiveness of GPS.

GPS versus Galileo: Balancing for Position in Space was originally written as a master's thesis for the Air University's School of Advanced Air and Space Studies (SAASS) at Maxwell AFB, Alabama, in June 2004. The College of Aerospace Doctrine, Research and Education (CADRE) is pleased to publish this SAASS research as a CADRE Paper and thereby make it available to a wider audience within the US Air Force and beyond.

DANIEL R. MORTENSEN
Chief of Research
Airpower Research Institute, CADRE

About the Author

Lt Col Scott W. Beidleman (BS, Pennsylvania State University; MS, University of Colorado; Master of Military Operational Art and Science, Air Command and Staff College; Master of Airpower Art and Science, USAF School of Advanced Air and Space Studies) is assigned to the Air Staff Directorate of Operational Plans and Joint Matters, Pentagon, Washington, DC. An experienced space operator, he earned his commission from the Air Force Reserve Officer Training Corps at Pennsylvania State University in 1988. Graduating from Undergraduate Space Training, Vandenberg AFB, California, in 1989, he went on to serve in a number of space-operations positions, including satellite mission-analysis officer, chief of standardization and evaluation, chief of operations training, chief of space-control war plans, space-surveillance crew commander, and operations officer. He has worked with a variety of space systems including the global positioning system at Schriever AFB, Colorado; the Deep Space Tracking System and the Low Altitude Space Surveillance System at Royal Air Force Feltwell, United Kingdom; and the AN/FPS-85 phased-array surveillance radar at Eglin AFB, Florida. He is a distinguished graduate of the Air Command and Staff College and a June 2004 graduate of the School of Advanced Air and Space Studies (SAASS), both based at Maxwell AFB, Alabama. Colonel Beidleman was selected to command the 533d Training Squadron, Vandenberg AFB, California, in the summer of 2006. His paper was recently recognized as the best SAASS thesis in airpower and technology for 2004.

Acknowledgments

I am indebted to Dr. Everett Dolman for suggesting a great thesis topic, providing guidance to strengthen my argument, and applying his outstanding editing skills to increase the quality of my written work. I also wish to thank Col Jon Kimminau for his insight in making a good product even better. Finally, and above all else, I thank my wonderful wife and daughter for enduring many "husbandless/fatherless" weekends and for those times when I was with them, but my thoughts were in "thesis-land." Their love and encouragement made this work possible.

Chapter 1

Introduction

And who can doubt that it will lead to the worst disorders when minds created free by God are compelled to submit slavishly to an outside will? When we are told to deny our senses and subject them to the whim of others?

—Galileo Galilei

In 1633 the Roman Catholic Church declared Galileo Galilei a heretic because his beliefs conflicted with the status quo.[1] Almost four centuries later, Europeans have christened their proposed global navigation satellite system (GNSS) with the independent thinker's name, a not so subtle challenge to the status quo dominated by America's global positioning system (GPS). Considering that GPS has become a global public good, an international utility paid for by the United States and free for use by anyone, and that most of Western Europe has been a staunch American ally since World War II, Europe's pursuit of the Galileo GNSS approaches heresy from an American perspective. Europe has broken ranks and is acquiring an independent space capability in a way that seems sure to conflict with American national interests.

In the post–Cold War environment, Europe has increasingly shown a desire to act independently of the United States to enhance its prestige and sovereignty. Despite long-standing cooperation agreements such as the North Atlantic Treaty Organization (NATO), Europe has pursued its own security initiatives, including the European Security and Defense Policy (ESDP) and the Rapid Reaction Force.[2] In this context, Galileo not only could strengthen European military independence, but also could bolster the European space program—adding credibility and prestige to Europe's effort to grow as a world power. Additionally, Galileo could challenge the US monopoly in the GNSS market and compete for its lucrative applications (air traffic control, shipping, etc.). This effort is not unprecedented—similar attempts to introduce pan-European competition in the past include the development of Airbus aircraft and Ariane launch boosters. Those efforts were seen as crucial to maintaining Europe's place in military matters and the most lucrative world markets. Competition with GPS is a challenge at least on par with these previous ventures and could prove even more rewarding.

Over the past quarter century, GPS has established itself as the world's standard for position, velocity, and timing information, providing a free, continuous, and all-weather navigation service to the entire planet. With innumerable applications such as guiding precision munitions, synchronizing the Internet, or locating a seafood restaurant in an unfamiliar city, GPS has become embedded in global society. Moreover, the United States openly shares technical details of the system's signal structure. Public documents specify the format of various data streams emanating from the satellites—data streams a receiver must recognize and decode to operate navigation and synchronization applications properly.[3] In this way, the United States provides key information enabling all interested parties to prosper by developing and marketing their own versions of GPS receivers. Finally, GPS is backed by the US government and operated by the US Air Force; clearly, the system's host is an extremely stable and competent authority.

Consequently, a puzzle arises: why is Europe pursuing the development of Galileo when a global space-based radio navigation system already exists that is free to all? Despite the high costs of developing and deploying its own redundant system, Europe is pressing ahead. From this action, follow-on questions emerge. Does GPS have deficiencies that Galileo will fix or improve? Are there motives that have not yet been made public? What are the implications of the proposed Galileo system for the United States? How should the United States respond?

To address these questions, I examined technical design documents, publications, and discourse from the European Union (EU) and the European Space Agency (ESA); various periodicals; and newspapers. I conducted my research in the midst of ongoing negotiations between the United States and the EU as they attempted to forge a cooperative agreement ensuring compatibility and interoperability between Galileo and GPS. While future talks may affect the relevance of the analysis contained herein, this study utilized data accurate as of 1 March 2004. Accordingly, I interviewed US military personnel and representatives from the US Departments of State and Transportation and attempted to do the same with corresponding EU officials. The sensitivity of these negotiations understandably tempered the candidness of some US government officials and resulted in no response from representatives of the European community with whom I inquired. I gleaned the European perspective chiefly from official government publications and press releases, promotional material from Galileo developers, and foreign newspapers and periodicals. In general, the bulk of the analysis relies on various defense- and space-related journals and periodicals to piece together the whole story.

My research shows that although GPS and Galileo were founded on different principles and were designed to meet the needs of different user communities, the two systems employ similar infrastructures and operational concepts. The key finding is that when operational, Galileo will provide better performance for global civilian users until GPS upgrades take effect. This overlap represents a window of opportunity for Europeans to take advantage of lagging GPS updates and seize a significant market share. After this, the two systems will provide analogous free services with comparable performance; however, only Galileo will offer a service guarantee for fee-paying customers.

Beyond providing an improved source for civil navigation (albeit temporary), Europe is pursuing Galileo to achieve a degree of independence from the United States. Trusting that satellite navigation will become increasingly embedded in the daily lives of its citizens, Europe views a public good controlled by a foreign power's military as a breach of sovereignty. Europe may also be acting on the belief that the prestige-enhancing aura of large space programs like Galileo will enrich its international standing. Besides political independence, Galileo will figure prominently in European efforts to develop a security apparatus independent of

INTRODUCTION

NATO, in part to protect against the possibility that the United States would degrade or deny GPS signals during a crisis. Lastly, Europe hopes Galileo will cultivate European technological independence by nurturing homegrown technical know-how in space technology that enables the EU's industrial capacity to compete with that of the United States on an equal footing.

In addition to strengthening European independence, the window of opportunity Galileo offers may include more than just the possibility of seizing a significant share of the satellite-navigation market from the United States. It opens the possibility that Europe could set a new global standard for navigation. With incredible growth forecasted for the satellite-navigation sector, if the EU can make itself the perceived leader in GNSS technologies and applications, it stands to gain considerably by overcoming the current US monopoly. Whether Galileo becomes—or GPS remains—the top satellite navigation service, either situation will have far-reaching effects. While a clear winner in the coming struggle for GNSS superiority is unknown, I argue here that civil and commercial users would reap the greatest benefit from the combination of both systems working together in a seamless GNSS.

The struggle will play out. The advent of Galileo presents a number of national security and economic implications for the United States. As originally proposed, Galileo would impede US space superiority by interfering with GPS signals and greatly complicating the ability to deny satellite navigation to hostile users. Economically, Galileo challenges US dominance of satellite navigation and poses a threat to usurp GPS as the world standard. In view of this potential reality, US concerns include ensuring fair trade and assured access to the global satellite-navigation market.

In response to early Galileo proposals, the United States initially downplayed the need for Galileo and took measures to forestall its development. As it became clear that Galileo would be developed over its objections, the United States changed its adversarial stance and sought ways to limit the potentially detrimental impact of Galileo on GPS users. Accordingly, the United States and EU continue to negotiate a cooperative agreement to produce an interoperable and compatible system for global navigation. To maintain and enhance its position, the United States must cooperate where it can and compete where it must by continuing efforts to develop a common standard for satellite navigation and by taking steps to strengthen the commercial and military competitiveness of GPS.[4] Specifically, I recommend that the United States and EU work towards standardized formats for satellite navigation, much like the standardization of Internet protocol, and that the United States formally separate the civilian and military aspects of GPS. These recommendations assume that Galileo will progress towards full operational capability as planned with no major delays and, when operational, will provide services as prescribed by the EU and ESA.

Overview

The analysis begins in chapter 2 with an assessment of the GPS and Galileo systems through a comparable evaluation of their respective strengths and weaknesses. Specifically, I examine the origins of the systems, their space and control segments, the services and capabilities provided, and the limitations and vulnerabilities. The chapter provides the reader with an understanding of both systems and determines if—and to what extent—Galileo provides a better source of navigation than GPS.

In chapter 3, I put forward a number of potential motives that, in combination, propelled Europe to build a GPS competitor. European incentive for an independent GNSS revolved around a desire for improved performance, independence from the United States, and economic aggrandizement. This requires an examination of performance in terms of accuracy, reliability, and vulnerability. I then explore European independence from political, security, and technological perspectives. The chapter concludes with a discussion of Europe's intent to increase its share of the potentially lucrative satellite-navigation market and Galileo's economic window of opportunity.

In chapter 4, I conclude the paper by examining the implications of Galileo from a US perspective, reviewing US policy, and recommending actions for the future. Galileo has potentially severe national security and economic implications, including encroachment on US space superiority and the potential loss of the GNSS market share. How the United Sates addresses Galileo's impact on US national interests in space sets the stage for future cooperation and confrontation on space policy as more nations attempt to become space powers.

Before settling into an analysis of the two space-based navigation systems, Europe's motives for Galileo, and the subsequent implications, it is necessary to put the issues into context. I begin with a concise history of space navigation and an overview of fundamental concepts.

Brief History of Navigation from Space

The idea of space-based navigation emerged from military necessity. While the terrestrial-based radio-navigation methods of the 1940s and 1950s supported intercontinental bombers in their missions to find city-sized targets, intercontinental ballistic missiles (ICBM) moved too fast to conduct the required navigational computations.[5] Developments in inertial navigation systems (INS) partially remedied the problem, but accuracy in INS without external updates decreased over time. In addition to missile guidance, as the United States developed the Polaris submarine-launched ballistic missile (SLBM) program, the Navy also needed an accurate method to determine the location of the submarine. To calculate a precise SLBM trajectory, one needs to know the location of the mobile launch site (submarine) as well as the target. "No INS alone would suffice for SLBM guidance. The drift over time would produce a navigational error too great to ensure target destruction."[6] Furthermore, terrestrial navigation systems lacked truly global coverage and were vulnerable to enemy attack. The solution lay in space.

Shortly after the launch of Sputnik, US researchers tracked the Soviet artificial moon by measuring the Doppler shifts of its frequency.[7] Using the same concept, researchers convinced the Navy to field a constellation of satellites broadcasting Sputnik-like signals. If the Navy knew the orbital locations of the satellites, its submarines could quickly determine their exact locations via the Doppler shifts. With this logic as a mandate, the

United States launched the Transit program, the world's first operational satellite-navigation system.[8] Transit satellites transmitted their orbital positions every two minutes on two signals, affording two-dimensional accuracies up to 25 meters for users tracking both signals.[9] The Transit constellation served military and civilian users quite well for 32 years and played an essential role supporting the sea-based leg of the nuclear triad.[10] However, a two-dimensional fix did not support aircraft or munitions in flight.

In the early 1970s, plans for a GPS emerged from the lessons of Transit and a combination of the Navy's Timation program and the Air Force's System 621B, with a goal of obtaining "greater accuracy with air-launched weapons and weapon systems."[11] Timation tested spaceborne atomic clocks, while System 621B demonstrated a new satellite-ranging signal based on pseudorandom noise (PRN).[12] When combined, these technologies and concepts offered the designs and techniques required to provide three-dimensional position, velocity, and time-transfer information.

Fundamentals of Navigation from Space

The new means for space-based navigation went well beyond measuring Doppler shifts and remain the fundamental principles of GPS and Galileo today. The operation is somewhat complex, but application has become essentially transparent and instantaneous to users. In order to locate one's position in three-dimensional space, a user needs eight ingredients from GPS at any given time: the distances or ranges between the user and at least four satellites, as well as the positions of these four satellites in space.

First, the user's GPS receiver measures the distance between itself and a satellite by measuring the time it takes for the signal to traverse from the satellite to the receiver's antenna.[13] The receiver measures the transit time by matching two codes. The process is straightforward. Each satellite transmits a unique PRN code (a pulse-train or stream of ones and zeroes) on its signal, allowing receivers to distinguish between satellites. Receivers internally generate the same codes. Since receivers and

INTRODUCTION

satellites are synchronized to the same time reference (GPS time), the satellite code and the receiver code should be identical. However, when the user set receives the satellite's code and compares it to its own code, the satellite code appears shifted. The receiver then slews its code until the ones and zeroes align. The amount of slewing required to match the codes represents the transit time of the satellite signal and, subsequently, the distance between the satellite and receiver, since the signal moves at the speed of light. In this manner the user obtains the ranges to four different satellites.

Second, each GPS satellite constantly transmits its unique orbital parameters in a data stream known as the navigation message (contained within the signal and the PRN code). Having locked onto a satellite's signal via the PRN code, the receiver downloads the satellite's position in space. Essentially, the satellite broadcasts, "I am *here*, and the time is (current time)." Once in possession of a set of four satellite positions and four ranges, the receiver calculates the user's location by solving four equations for four unknowns: the user's altitude, longitude, and latitude, together with the receiver's clock error.[14] The accuracy of this solution depends heavily on which four satellites the receiver chooses.

The accuracy of a user's position fix depends on where the satellites are in the sky with respect to the user, otherwise known as satellite geometry. Consider the geometry of a polyhedron with the user at the vertex (fig. 1). The volume of the polyhedron affects GPS accuracy. As a rule, greater satellite spacing results in a larger volume and better accuracy. Ideally, the best geometry results from having one satellite directly overhead and three satellites spread equally near the user's horizon. With as many as 12 GPS satellites in view at any given time, the receiver must carefully select the optimum satellite combination to obtain the best fix.[15] Again, this is a relatively instantaneous and transparent function of the receiver's computer. The receiver examines all possible satellite combinations and chooses the solution with the best satellite geometry. The user is now in receipt of three-dimensional position information accurate to within 50 feet.[16]

INTRODUCTION

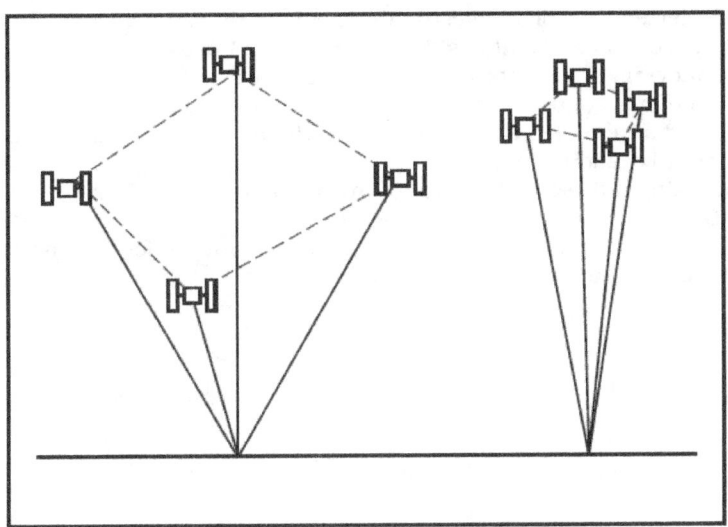

Figure 1. Satellite geometry

Now that I have described the issue to be analyzed, accounted for the method of analysis and the outline of the thesis, and reviewed the history and fundamentals of space-based navigation, the reader is equipped to embark on a comparative analysis of American and European satellite-navigation systems.

Notes

(Notes for this chapter and the following chapters appear in shortened form. For full details, see the appropriate entries in the bibliography.)

1. Halsall, "Crime of Galileo."
2. "EU Law + Policy Overview."
3. Warner, "GPS ICD-200," 1.
4. Parkinson, "Capability and Management Issues."
5. Rip and Hasik, *Precision Revolution*, 57.
6. Ibid.
7. Ibid., 60.
8. Federation of American Scientists, "Military Space Programs."
9. Ibid.
10. The Navy deactivated Transit in December 1996.

INTRODUCTION

11. Sietzen, "Galileo Takes on GPS," 38.
12. Pace et al., *Global Positioning System*, 239.
13. Moving at the speed of light, the signal's travel time is roughly one-eleventh of a second.
14. For a more detailed discussion of the navigation solution, see chaps. 1 and 2 of Logsdon, *Navstar*.
15. Using a five-degree elevation mask, according to European Commission, *Galileo: Mission High Level Definition*, 28.
16. Logsdon, *Navstar*, 64. The quoted accuracy refers to the Standard Positioning Service.

Chapter 2

GPS versus Galileo

GPS has become a global utility.
—Pres. Bill Clinton
1 May 2000

In this chapter I compare the two navigation systems to determine critical similarities and differences and to evaluate their strengths and weaknesses. Specifically, I attempt to determine GPS operational deficiencies from a European perspective and the extent to which Galileo intends to improve upon them. The assessment begins with a review of the primary purpose and sponsorship of each system, followed by an examination of active and proposed system infrastructures, services, limitations, and vulnerabilities.

Purpose and Sponsorship

The raison d'être and sponsorship of the two navigation systems produce two fundamentally different outlooks. As previously discussed, GPS was driven by the military's need for increased weapon's accuracy. Consequently, the US Air Force owned and operated the system—fully funded by the US taxpayer—and the Department of Defense (DOD) maintained ultimate authority. Recognizing the growing use and importance of GPS to the civilian community, the US government established the Interagency GPS Executive Board (IGEB) in 1996. While the Air Force still operates the system, the IGEB manages senior-level policy for GPS and is chaired jointly by the DOD and Department of Transportation (DOT). Nevertheless, and regardless of its dual-use potential, the primary purpose of GPS is "enhancing the effectiveness of US and allied military forces."[1] US national security remains the top policy goal for decisions concerning GPS.[2]

In contrast, Galileo came about as a direct response to the perception that GPS catered to US military requirements at the expense of global civilian needs. Although the United States promoted GPS as a worldwide utility, it did not promote global participation in managing that increasingly depended-upon global resource, nor would it guarantee continued access to it. Thus, Galileo emerged as a joint venture between the European Commission (EC) and the ESA, spurred by transport ministers with a decidedly nonmilitary perspective. Furthermore, Galileo is funded through a public-private partnership in which the EC and ESA provide funding in tandem with private companies participating in the project. When operational, a consortium-created private company, referred to as the Galileo Operating Company (GOC), will operate and maintain the constellation. In short, "the US places priority on security and allied military capability, [and] Europe places it on commercial viability."[3]

System Infrastructure

Both the GPS and Galileo systems are subdivided into three components: the space segment, comprised of the satellites; the control (or ground) segment, consisting of the command and control infrastructure; and the user segment, encompassing the end user or customer.

Space Segment

The purpose of the GPS space segment is to transmit timing pulses and satellite positional data to users worldwide. The GPS space segment is comprised of 24 satellites in a "Walker constellation" at an altitude of 10,898 nautical miles (roughly 20,200 km), organized in six orbital planes equally spaced in right ascension around the earth, with an inclination of 55 degrees.[4] Walker constellations (named after the British Royal Aircraft Establishment's J. G. Walker) are satellites configured in circular orbits with common altitudes and inclinations that provide global coverage of the earth.[5] The design of the GPS constellation guarantees that at least five satellites with favorable satellite geometry (table 1) are always in view to users worldwide to meet accuracy requirements.[6] Since their inception, the 2,175-pound GPS satellites have been launched individually from Cape Canaveral, Florida, on Delta II boosters.[7]

Table 1. Number of visible satellites for various masking angles

Receiver elevation masking angle	Number of visible Galileo satellites	Number of visible GPS satellites	Total
5°	13	12	25
10°	11	10	21
15°	9	8	17

Source: European Commission, *Galileo–Mission High Level Definition*, Version 3.0, 23 September 2002.

Four generations of GPS satellites have served the mission thus far—Block I, Block II, Block IIA, and Block IIR (replenishment). The Block I satellites were prototypes to test the concept of navigation from space. Block II vehicles, the first operational series, added radiation hardening and a 14-day autonomous navigation message to increase survivability during war. Further emphasizing military utility, Block II satellites also debuted selective availability and antispoofing. With selective availability the United States can degrade GPS accuracy to unauthorized users. Similarly, antispoofing allows the United States to deny high-accuracy GPS signals to real and potential enemies through encryption and prevents enemies from transmitting false GPS-like signals intended to fool or corrupt GPS receivers. The Block IIA series extended the autonomous navigation message to 180 days, providing slowly degrading data for six months in the event the ground-control segment was destroyed. The most recent Block IIR satellites added additional radiation hardening and operational redundancy, as well as a cross-link ranging mode that enables IIR vehicles to update their own navigation message without support from the ground for up to 180 days. The current constellation is comprised of Block II/IIA and IIR vehicles. Future satellite generations include Blocks IIR-M (modified), IIF (follow-on), and GPS III. These future system upgrades will add additional signals for civilian and military users and increase signal power. In particular, the GPS III constellation "will have the ability to surge the signal over a specific area for certain intervals" via spot beams.[8]

Figure 2. GPS satellite (Courtesy of ESA at their Web site at http://www.esa.int/externals/ images/estec-photo-archive/1059.jpg.)

GPS satellites (fig. 2) transmit their timing pulses and positional data to Earth via radio waves in the L-band frequency portion of the electromagnetic spectrum. The system currently uses two carrier signals, known as L1 (at 1575.42 MHz) and L2 (at 1227.6 MHz). The timing pulses (i.e., the unique PRN codes) are superimposed on the carriers, and the navigation message is superimposed on the timing-pulse trains.[9] In addition to specific positional information, the navigation message of each satellite also carries satellite health status and an almanac listing the orbital positions of the entire constellation. Furthermore, GPS employs pulse trains, or PRN codes, in two different formats: the precision (P)-code and coarse acquisition (C/A)-code. The P-code resides on L1 and L2 and is available only to authorized users, mostly military and government. The C/A-code resides only on the L1 signal and is available to everyone. As designed, the P-code has two primary advantages over the C/A-code. First, the P-code provides a more precise fix because of its faster chipping rate. It streams down at 10 million bits per second (bps) compared to the C/A-code chipping rate of one million bps.[10] Second, it further boosts accuracy as it's transmitted on both (L1 and L2) signals at different frequencies, enabling users to correct for range errors due to the ionosphere (occurring in the atmosphere where propagation of radio waves is hampered due to ionization of gases). Hence, the GPS design segregates the user pool into the haves (military users) and have-nots (civilian users) regarding precision accuracy. Finally, powered by solar arrays generating 700 watts, the satellites transmit signals at low power to Earth.[11] In fact, GPS signals are so weak they have been likened to a "whisper at a cocktail party," leading to vulnerabilities discussed later.[12]

Similar to GPS in its operations, the proposed Galileo space segment will perform the space navigation mission with only minor differences. Galileo employs more satellites in fewer orbital planes with a slightly higher altitude and inclination. Specifically, Galileo consists of 30 satellites in a Walker constellation at an altitude of 23,616 km, equally spaced within three orbital planes with a 56-degree inclination. The higher altitude and inclination afford Galileo better coverage in the higher latitudes, including some polar regions, than does GPS—especially benefiting civilian users in Scandinavian countries. Moreover, the higher number of satellites increases the availability of satellites visible to a user, thereby improving satellite geometry and enabling better accuracy. Of note, although Galileo's additional satellites only marginally improve satellite geometry compared to GPS, combining the two systems would produce a significant increase in visible satellites (see table 1). Interestingly, Boeing—the GPS Block IIF contractor—has advocated switching GPS to a Galileo-style 30-satellite, three-plane constellation in the future.[13]

Physically, Galileo satellites are smaller and lighter than their GPS counterparts (fig. 3). Unhindered by military-threat requirements, Galileo satellites will forgo nuclear hardening, will not carry a nuclear-detonation-detection payload, and do not require a six-month autonomous operational capability. Hence, the Galileo

Figure 3. Galileo satellite (Courtesy of ESA at their Web site at http://esamulti media.esa.int/images/navigation/galileo02775A4.jpg.)

spacecraft, at approximately 650 kg, will weigh roughly half as much as GPS. Consequently, Galileo's smaller size supports the launch of multiple satellites aboard a single European Ariane booster to quickly populate the original constellation, with smaller launch vehicles envisioned to replace failed satellites.[14]

Similar to GPS, Galileo will transmit its timing and navigation information in the L-band spectrum. However, whereas GPS currently provides only two signals, Galileo will provide 10 navigation signals to support a number of different services. Specifically, Galileo plans to employ two signals on the E5A band centered at 1176.45 MHz, two signals on E5B at 1207.14 MHz, three signals on E6 at 1278.75 MHz, and three signals on E2-L1-E1 at 1575.42 MHz.[15] This proposed signal plan has stirred considerable controversy regarding potential interference with existing GPS signals (an issue discussed in more depth in chap. 4). Also similar to GPS, Galileo signals will carry different PRN code schemes that effectively segregate users into three distinct groups: the general public, commercial users, and authorized government users.[16] However, unlike GPS, Galileo will transmit the PRN code available to the general public on two signals, enabling everyone to correct for ionospheric delays and obtain higher accuracy. In essence, GPS segregates users by controlling access to better accuracy, whereas Galileo divides users by controlling access to ancillary data. One primary type of segregated ancillary data is integrity, which constitutes a major difference between GPS and Galileo. The EC defines integrity as the ability "to provide timely warnings to the user when [the system] fails to meet certain margins of accuracy."[17] Thus, Galileo plans to constantly monitor system accuracy and quickly update the constellation upon detection of a problem.[18] Specifically, Galileo will broadcast integrity flags in the navigation message with a time-to-alert of six to 10 seconds.[19] Currently, GPS can take as long as 30 minutes to notify users of an out-of-tolerance condition.[20] Galileo will be the first GNSS to incorporate real-time signal-integrity monitoring, a capability not planned for GPS until the GPS III upgrade, currently scheduled for 2012.[21]

Control Segment

As with the GPS and Galileo space segments, the control segments of the two systems are very similar. While the satellites of the space segments constantly transmit their locations to the users, the satellites themselves do not know where they are. They transmit only what they are instructed to transmit. The control segment on the ground develops, monitors, and updates each satellite's navigation message and then feeds the data to the satellite for retransmission.

The US Air Force maintains and operates the GPS constellation via a control segment comprised of the Master Control Station (MCS), monitoring stations, and ground antennas. The MCS, located at Schriever AFB, serves as the central processing facility for GPS. Operated continuously by Air Force crew personnel, the MCS houses the operations center for command and control of the constellation, the computers used to

GPS VERSUS GALILEO

predict orbits, and the array of atomic clocks that constitutes the system's timing reference, known as GPS time.

Linked to the MCS, five unmanned monitoring stations support GPS, spread across the globe at Hawaii, Colorado Springs, Ascension Island (South Atlantic Ocean), Diego Garcia (Indian Ocean), and Kwajalein (western Pacific Ocean). The monitoring stations constantly track each satellite in view, measure the range to each satellite (known as a *pseudorange*), and download each navigation message. Then the stations send the pseudoranges and navigation messages to the MCS for processing.

The MCS receives the constant flow of information from the monitoring stations and calculates a fresh predicted orbit (ephemeris) for each satellite from the pseudoranges. Additionally, the MCS examines the navigation message of each satellite to verify that the satellites are transmitting the correct ones and zeroes. Finally, the MCS updates the navigation message for each satellite based on the new orbit predictions and sends the resulting navigation uploads to one of five ground antennas.

The GPS control segment employs four dedicated, unmanned ground antennas located at Cape Canaveral, Diego Garcia, Kwajalein, and Ascension Island. A fifth antenna at Schriever AFB is also available upon request to support GPS requirements.[22] The ground antennas provide the MCS an uplink and downlink capability. They receive navigation uploads and other commands from the MCS and transmit them to the satellites. Moreover, they collect and transmit telemetry, tracking, and commanding (TT&C) data, enabling the MCS to monitor and maintain the constellation's health and to control each satellite. Thus, the GPS control segment is a data-processing loop that maintains a continuous flow of accurate navigation information.

The Galileo control segment greatly resembles the GPS infrastructure. As proposed, a private business enterprise to be called the GOC will manage the constellation from two navigation system control centers (NSCC) located somewhere in Europe (possibly France and Germany, based on their sizeable contributions), along with a global network of unmanned orbitography and synchronization stations (OSS) and TT&C stations.[23]

Like the GPS MCS, the Galileo NSCC serves as the "heart of the system and includes all control and processing facilities," providing orbit determination and maintaining time synchronization (with Galileo time as the reference).[24] As with GPS monitoring stations, the Galileo OSS collects and measures navigation data and passes it to the NSCC. Finally, similar to GPS ground antennas, Galileo TT&C stations provide uplink and downlink capabilities, linking the NSCC with the constellation. Despite all of these resemblances, the Galileo control segment is not a mirror image of its GPS counterpart.

The major difference between the control segments is Galileo's addition of integrity monitoring. Galileo satellites are designed to provide integrity alerts to users within the navigation signal-in-space. Thus, the Galileo control segment includes an Integrity Determination System comprised of integrity monitoring and uplink stations.[25] The system will monitor each satellite's signals, determine whether the signals are outside specifications, and uplink an integrity flag to the constellation identifying faulty satellites. Hence, a user's receiver will reject the signals from satellites identified in the alert.[26]

User Segment

Since the DOD developed GPS to support national security, the US armed forces are the primary intended customer. However, in 1983 after the Soviets shot down a Korean airliner that erroneously wandered into Soviet airspace, Pres. Ronald Reagan declared that the United States would provide civilian airliners access to GPS signals, essentially establishing GPS as a dual-use system.[27] Pres. Bill Clinton reiterated the US commitment to providing civilian access free of charge in the 1996 Presidential Decision Directive NSTC-6.[28] In contrast, Europe has marketed Galileo as a public GNSS geared to civilian and commercial user requirements and has downplayed Galileo's military utility. Generally speaking, nonmilitary customers comprise the overwhelming majority of all GNSS users; however, GPS places the military user above the civilian for reasons of national security.

Services

GPS provides position, navigation, and timing (PNT) services with two different levels of accuracy: the Standard Positioning Service (SPS) and the Precise Positioning Service (PPS). The unencrypted SPS offers PNT services free of charge to all users. Based on the less accurate C/A-code transmitted on L1, the SPS cannot self-correct for ionospheric errors and so produces less accurate navigation solutions than the PPS. Additionally, to prevent adversary use of GPS against US forces during conflicts, the US government can intentionally degrade SPS accuracy via selective availability. For many years, selective availability fixed SPS accuracy at roughly 100 meters. However, in 2000 President Clinton directed that the DOD turn off selective availability; consequently, SPS accuracy increased tenfold to roughly within 10–20 meters.[29] While the SPS is unencrypted, the United States currently encrypts the PPS and restricts access to authorized users, mainly the armed forces and government agencies. Based on the P-code transmitted on L1 and L2, the PPS offers more accurate PNT services with positional accuracy of approximately 25 feet.[30]

In contrast to GPS, Galileo plans to offer five types of services—Open Service (OS), Commercial Service (CS), Safety-of-Life (SoL) Service, Public Regulated Service (PRS), and Search and Rescue (SAR) Support Service.

Open Service

OS is similar to the GPS SPS in that it is intended for the general public and is provided for free. However, since OS will be transmitted on two frequencies, users of this basic service can correct for ionospheric effects and obtain better accuracy than with the GPS SPS. Specifically, the ESA expects to achieve four-meter accuracy with a service availability of 99.8 percent.[31] GPS will not provide this level of accuracy until the Block IIF constellation is operational circa 2012.[32] Again, similar to the GPS SPS, the OS provides no service guarantee or integrity information to the general public. Like the GPS's civilian users, Galileo's OS customers use Galileo at their own risk.

Commercial Service

CS is a combination of OS plus two encrypted signals separated in frequency from OS signals.[33] Like OS, CS will not explicitly carry integrity data; however, CS accuracies will be guaranteed. Designed to support users requiring higher performance than OS, CS's additional signals allow the development of professional applications such as producing high-data-rate broadcasting, resolving ambiguities in differential applications, and integrating Galileo with wireless communications.[34] Access to the encrypted signals will be restricted to fee-paying users who will subscribe to CS. Third-party service providers will determine the specific services offered and will purchase the rights to utilize the encrypted signals via a license agreement with the GOC.[35] Finally, the company will provide a guarantee for disruption or degradation of service and will provide timely warning to users. Failure to meet standards would lead to compensation to affected users and/or service providers. Service guarantees addressing liability constitute a major difference between Galileo and GPS.

Safety-of-Life Service

SoL will offer the same accuracies as OS, but it will provide both integrity data and service guarantees for a fee. SoL is designed to serve safety-critical users who require precision accuracy and signal reliability. Anticipated customers include airlines, trains, and transoceanic maritime companies.[36] With SoL, Galileo plans to comply with "levels of service stipulated by law in various international transportation fields," such as those prescribed by the International Civil Aviation Organization.[37] SoL's integrity monitoring is essential to meeting this goal. Galileo will reportedly inform users of out-of-tolerance conditions within six to 10 seconds of occurrence, supporting safety-critical applications such as Category I landings (aircraft landings with weather conditions of a 200-foot ceiling and visibility of one-half mile). However, this additional service comes at a price. Users will need specialized receivers to get the enhanced signals,[38] and the EC retains the option to encrypt integrity data and administer access fees.[39]

Public Regulated Service

The objective of PRS, according to the EC and ESA, "is to improve the probability of continuous availability of the SIS [signal-in-space], in [the] presence of interfering threats."[40] Envisioned as a protected navigation service for government and public-service users, PRS will employ robust signals with interference mitigation technologies to reduce susceptibility to jamming and interference from terrorists, criminals, or hostile entities that could affect national security.[41] Furthermore, PRS must remain operational during crises, when other services may be jammed. Hence, Galileo transmits PRS on two wideband signals (to increase jamming resistance) and spectrally separates them from other Galileo services, so these other services "can be denied without affecting PRS operations."[42] Additionally, PRS will be encrypted to restrict access to interference-mitigation technologies and to prevent hostile use of PRS against EC member states (in this paper, the word *states* used in context with Europe refers to EU nation-states).[43] Accordingly, EC member states will control PRS access via cryptological keying systems. Based on this description, the Galileo PRS sounds very similar to GPS PPS and presents a potential military capability in a system strictly trumpeted as "the first satellite . . . navigation system specifically for civil purposes."[44] In fact, while Galileo is frequently touted as "a civil system, operated under public control" and "a non-military programme," the design and spectral locations of PRS signals mirror future GPS military upgrades, potentially conflicting with US navigation warfare (NAVWAR) concepts (see chap. 4).[45]

Search and Rescue Support Service

SAR will augment the Cospas-Sarsat system, which assists international search and rescue efforts by detecting and locating distress signals worldwide. Galileo satellites will employ SAR transponders that will detect distress alerts and relay the detection to Cospas-Sarsat ground stations. Moreover, Galileo will also send an acknowledgement to the stranded persons informing them they have been located. Consequently, Galileo will reportedly fine-tune alert-location accuracy, greatly improving the current specification of five kilometers.[46] Also,

Galileo will provide near-real-time reception of distress messages, greatly reducing the current wait time of one hour.[47]

In short, both GPS and Galileo provide basic PNT services open to all users, as well as augmented services restricted to authorized users. However, Galileo plans to offer additional features such as service guarantees, global-integrity monitoring, and additional data services supporting commercial markets in an attempt to capitalize on GPS limitations from a civilian perspective.

Limitations and Vulnerabilities

In general, the performance of GPS and the impact of its PNT capabilities have led to its perception as a global utility.[48] However, like every system, GPS has limitations and vulnerabilities. While the proposed Galileo design will purportedly overcome several GPS deficiencies, including liability, integrity, and inadequate civilian accuracy, other issues affect both systems. These include "urban canyons" (often occurs in cities, created by tall structures obscuring signals), susceptibility to jamming, and hostile use by potential adversaries.

One of the primary differences between GPS and Galileo is the latter's liability service guarantee. Unlike GPS, Galileo plans to provide a guarantee against disruption of service in terms of accuracy, continuity of availability, and integrity, where interruptions "would have significant [safety] or economic impacts."[49] This translates to a service guarantee for Galileo's CS and SoL service. Thus, Galileo will provide a legal framework to increase the confidence of users previously reluctant to utilize space-based radio navigation signals as a primary means of navigation. As mentioned before, the GOC (the private company chosen to manage the constellation) will commit to providing the signal quality required to support the specified services and will compensate users if signal quality falls short of specifications without adequate warning.[50] In this manner, Galileo users sidestep the potential barriers faced by GPS users who file claims against the US government as GPS owners—GPS does not provide a service guarantee.

On the surface, a service guarantee appears marginal or insignificant—the general public will largely ignore it. However, a significant subset of users (such as air traffic controllers) would highly value service guarantees for safety-critical or precision operations. The functions these users provide, primarily accomplished through national governments employing Galileo and GPS as the basis of their transport policies, do affect entire populations. The lack of a service guarantee could impede GPS's ability to compete in this critical niche. As a senior fellow on the Council on Foreign Relations observed, "Until GPS is certifiable for aviation use worldwide, its usefulness will be unavoidably curtailed."[51]

The viability of Galileo's service guarantee remains to be seen. Its credibility depends on the system's ability to compensate a user's loss, leading to a multitude of contractual and liability issues beyond the scope of this study. In short, the EU foresees the guarantee relying on legal mechanisms "to prevent, inform, alert, or compensate failure, disruption, or provision of a service" due to failing specifications.[52] These may include certification of risks, licensing usage, and mechanisms to manage compensation or reimbursement and jurisdiction/recourse issues, yet to be defined.[53] The crux of Galileo's service guarantee is adequate warning of substandard performance, accomplished via integrity monitoring.

Lack of real-time integrity monitoring is another shortcoming of GPS from the civilian perspective. Currently, if a GPS satellite's navigation signal drifts out of tolerance, the GPS control segment must schedule a contingency contact to refresh the satellite's memory with a new navigation upload or set the satellite's health flag within the navigation message. Depending on satellite visibility, this process could take up to 30 minutes before the satellite transmits corrected information.[54] Meanwhile, users are not warned of the out-of-tolerance condition and could continue to use less accurate data.[55] In contrast, Galileo plans to directly warn users of substandard performance in less than 10 seconds, allowing customers "of standard commercial services to react rapidly to malfunctions,"[56] albeit potentially for a fee. Similar to the reaction to Galileo's service guarantee, one might conclude that *integrity for a price* would appeal only to a small number of professional users, certainly not the majority of recreational GNSS users. Although integrity could be characterized as a niche service, many users who prefer integrity provide services that affect the mass public, such as the airline, banking, telecommunications, and transport industries. Therefore, while integrity may directly serve the needs of only a few niche users, integrity indirectly affects virtually everyone. Consequently, Galileo bests GPS by offering fee-based integrity. However, with respect to free/open services, neither GPS nor Galileo provides real-time integrity monitoring, though Galileo's free services will reportedly be more accurate.

A third shortcoming of GPS is its level of accuracy afforded to civilian users. This deficiency is in part a deliberate characteristic of the system. Designed as a national-security asset, GPS provides better accuracy for military users. Selective availability intentionally decreased civilian accuracy in the past, but even without selective availability, civilian accuracy is inherently less precise than Galileo's because the SPS is transmitted on only one signal. In contrast, the military utilizes two frequencies in order to improve accuracy by correcting for ionospheric errors. Galileo plans to transmit its free OS on dual frequencies, enabling higher accuracies for civilians than is currently possible with GPS.[57] In addition to dual-frequency use, the GPS constellation design causes accuracy to degrade at higher latitudes. Accuracy degrades because satellite geometry (as defined in chap. 1) diminishes at higher latitudes. At latitudes above 55 degrees, GPS satellite spacing becomes increasingly confined to the user's horizon, with no satellites directly overhead. Thus, poor satellite geometry results in lower overall accuracy. Consequently, Galileo hopes to improve satellite geometry at higher latitudes by orbiting satellites in higher inclinations and altitudes than those offered by GPS, thereby providing users with satellites higher on their horizon (although not directly overhead).

Lower accuracy at higher latitudes is one of the orbit-limiting characteristics of GPS. Accuracy also suffers in cities and other areas of severe occultation where signals are prohibited from reaching the user. In fact, such urban canyons can limit service availability to 55 percent of a typical city's territory.[58] While Galileo will also endure the same coverage problems on its own, the combination of 29 GPS and 30 Galileo satellites could improve positioning-service coverage to 95 percent of an urban area.[59]

In addition to signal blocking by natural and man-made barriers, jamming can also affect signal availability by denying service to local areas. By design, GPS signals reach the user at very low power levels of only a few milliwatts.[60] With such a weak signal, even low-watt jammers can be effective. Less than one watt of power will suffice to jam standard receivers at a range of 25 kilometers, while a 100-watt jammer "can blanket a 65-kilometer region."[61] To put this into contemporary perspective, "a single electronic jammer radiating at one-tenth of a watt could prevent [civilian receivers] from tracking GPS within the Baghdad metropolitan area."[62] In a real-world example, at a tank competition in August 2000 sponsored by the Greek government, a French security agency jammed British and US tanks during trial demonstrations, causing significant navigation problems.[63] Although Galileo plans to transmit at a slightly higher power, it too will be susceptible to intentional and possibly inadvertent jamming. Various techniques are available to reduce this susceptibility, but susceptibility to jamming is not altogether a bad thing because it enables the United States to deny locally the use of GPS to potentially hostile individuals, parties, etc.[64]

GPS and Galileo are mutually susceptible to hostile use by an enemy or adversary. Since both systems offer free and open services to anyone with a receiver, significantly increased accuracy is available to rogue states, terrorists, and rising peer competitors like China. Originally, selective availability served to deprive hostile users of precise GPS PNT services, but it also deprived legitimate civilian users, who outnumber military users 100 to one.[65] For this reason, the United States terminated the capability in 2000. Likewise, Galileo currently has no method of denying its OS to undesirable users (other than jamming). As a result, many states, including EU members and those in the so-called axis of evil, have enhanced their military capabilities via GPS.[66] North Korea has reportedly utilized GPS on its submarines, China is integrating it into fighter aircraft, and Iran Aircraft Manufacturing plans to incorporate GPS on board new variants of its Ababil unmanned air vehicle.[67]

This chapter has examined GPS and Galileo for similarities and differences to determine if GPS had deficiencies that Galileo could exploit and, ultimately, to forecast if Galileo could provide a better source for navigation than GPS. Although the two systems were founded on fundamentally different visions, the similarities outnumber the differences. GPS is designed to support national security, while Galileo is designed to make Euros. Nevertheless, the two systems share similar infrastructures, performing virtually the same functions to provide like PNT services.

While GPS has arguably become a global public utility, the system is not without significant limitations and vulnerabilities. Galileo claims it will resolve several GPS deficiencies, including liability, integrity, and inadequate civilian accuracy. Even if these assertions prove true, GPS modernization programs (see subsequent chapters) will eventually level the playing field once again. Moreover, GPS and Galileo are mutually deficient and vulnerable regarding urban canyons, jamming, and hostile use by potential adversaries.

Assuming Europe fields Galileo as claimed, Galileo will outperform GPS in the near term from a civilian perspective by providing a guaranteed service with better accuracy and global-integrity monitoring—for a price. However, in the not-too-distant future the United States will upgrade GPS, enabling it to provide the same services to the same users for free. Given this inevitable outcome, it remains uncertain why Europeans would expend the enormous capital involved to compete with a successful and reliable system not only backed by the US government but also available free of charge.

Notes

1. United States Mission to the European Union, "US Global Positioning System."
2. Ibid.
3. Julie Karner (assistant director, Office of Space and Advanced Technology, US Department of State), interview by the author, 5 Jan. 2004.
4. As of 14 Jan. 2004, GPS had 29 operational satellites in orbit. Maj Michael Brownworth, GPS Operations flight commander, interview by the author, 13 Jan. 2004.
5. Wertz and Larson, *Space Mission Analysis and Design*, 175.

6. Pace et al., *Global Positioning System*, 218.
7. The weight listed is the on-orbit weight of Block II satellites. Block IIR satellites weigh 2,370 pounds, and Block IIF satellites will weigh 3,407 pounds.
8. Sirak, "Holding the Higher Ground," 21.
9. Logsdon, *Navstar*, 20–21.
10. Ibid., 20.
11. The power listed is for Block II satellites. Block IIR satellites generate 1,136 watts, and IIF will generate 2,900 watts.
12. Divis, "This Is War."
13. Richardson, "GPS in the Shadows of Navwar," 26.
14. Benedicto et al., "Galileo: Satellite System," 9.
15. European Commission, *Galileo: Mission High Level Definition*, 41–42.
16. Ibid., 29.
17. Ibid., 16.
18. Galileo will provide integrity monitoring independent of augmentation systems like the European Geostationary Navigation Overlay Service (EGNOS). Also, Galileo will provide global-integrity monitoring, whereas EGNOS is a regional system serving greater Europe.
19. Benedicto et al., "Galileo: Satellite System," 13.
20. Brownworth, interview.
21. Pappas, "Effects of the Galileo," 8.
22. Unlike the dedicated antennas, the antenna at Schriever AFB (known as PIKE) is part of the Air Force Satellite Control Network (AFSCN) and supports many satellite programs in addition to GPS.
23. Benedicto et al., "Galileo: Satellite System," 1. The NSCC, OSS, and TT&C are also known as the Galileo Control Center, Galileo Sensor Station, and Galileo Uplink Station, respectively.
24. European Commission, *Galileo: Mission High Level Definition*, 30.
25. Current documentation implies separate equipment for integrity functions vice assigning the mission to OSS and TT&C stations.
26. Benedicto et al., "Galileo: Satellite System," 13.
27. Statement by the principal deputy press secretary to the president, 16 Sept. 1983, in Rip and Hasik, *Precision Revolution*, 429.
28. Presidential Decision Directive NSTC-6, *US Global Positioning System Policy*.
29. US Department of State, "US Global Positioning."
30. Logsdon, *Navstar*, 64.
31. European Commission, *Galileo: Mission High Level Definition*, 14.
32. Parkinson, "Capability and Management Issues."
33. Ibid., 15.
34. Wilson, *Galileo: The European Programme*, 20.
35. European Commission, *Galileo: Mission High Level Definition*, 15.
36. Wilson, *Galileo: The European Programme*, 20.
37. *Galileo: Mission High Level Definition*, 16.
38. Commission of the European Communities, *Commission Communication to the European Parliament*, 13.

39. Karner, interview.
40. European Commission, *Galileo: Mission High Level Definition*, 17.
41. Ibid.
42. Wilson, *Galileo: The European Programme*, 21.
43. The European Commission (EC) contains the executive function of the European Union (EU). The EU, which currently has 25 member "states" or nations, works towards and oversees the economic and political integration of these states.
44. Wilson, *Galileo: The European Programme*, 5.
45. European Commission, *Galileo: Mission High Level Definition*, 6; and European Parliament, *Report on the Commission Communication to the European Parliament*, 40.
46. European Commission, *Galileo: Mission High Level Definition*, 19.
47. Ibid.
48. Adams, "GPS Vulnerabilities," 11.
49. European Commission, *European Dependence*, sec. 3; and European Commission, *Galileo: Mission High Level Definition*, 11.
50. European Commission, *European Dependence*, 12.
51. Braunschvig et al., "Space Diplomacy," 159.
52. European Commission, *European Dependence*, 24.
53. Ibid.
54. Brownworth, interview.
55. Augmentation systems like the Wide Area Augmentation System (WAAS) provide localized or regional GPS integrity information to civilian users, but these systems operate independently of GPS. Galileo allegedly will inherently provide global integrity coverage.
56. European Commission, *European Dependence*, sec. 3.
57. Future GPS upgrades will provide additional signals for civilian users, but Galileo will provide this service first.
58. European Parliament, *Report on the Commission Communication to the European Parliament*, 28.
59. Ibid.
60. A milliwatt is approximately -160 dBW (power level in decibels relative to one watt). Alterman, "GPS Dependence," 54.
61. Rip and Hasik, *Precision Revolution*, 278; and Richardson, "GPS in the Shadows of Navwar," 22.
62. Rip and Hasik, *Precision Revolution*, 280.
63. Adams, "GPS Vulnerabilities," 12.
64. Jam-resistance techniques have fortified military receivers such that a 100-watt jammer would need to be within 0.3 nautical miles to cause significant problems. Roos, "Pair of Achilles Heels," 22.
65. GPS Joint Program Office, "Navstar GPS Fact Sheet."
66. Iraq, Iran, and North Korea were termed *axis of evil* nation-states in Pres. George W. Bush's State of the Union Address on 29 Jan. 2002.
67. Snyder, "Navigating the Pacific Rim"; and Adams, "GPS Vulnerabilities," 11. China is not exactly an axis of evil state, per se.

Chapter 3

Why Galileo?

If you use your parent's car, there will come a day when it's not available.

—Gilles Gantelet
European Commission Spokesman

In 1996 the US government pledged to provide GPS "for peaceful civil, commercial and scientific use on a continuous worldwide basis, free of direct user fees," and it has largely kept its word.[1] Since its operational inception in 1994, GPS remains omnipresent and complementary. However, Europe, America's traditional ally for the past six decades, has decided to expend 3.6 billion euros to pursue its own satellite-navigation system. Why would anyone pay to build a capability that is already available for free? Only the Soviet Union, America's Cold War enemy, saw fit to build a Global Navigation Satellite System (GLONASS) primarily to guide its bombers and missiles against America and its allies.[2] Europe harbors no such plans, so why Galileo?

In this chapter, I examine Europe's rationale to build a separate satellite-navigation system. I contend there are three main sources of motivation propelling Europe towards Galileo: improved performance, independence from the United States, and economic opportunity.

Performance

Many Europeans believe that "consumer reliance on satellite navigation will turn into dependence, as [its] use becomes an essential tool for business and daily lives."[3] They are further concerned that as business expands and reliance increases, GPS may not be upgraded to meet future needs in a timely fashion. Hence, GPS performance in the form of accuracy, reliability, and vulnerability has become a primary concern and motive for European development of Galileo.

Originally, the United States designed GPS as a military support system and intentionally degraded civilian-accessible accuracy through selective availability to approximately 100 meters. This policy continued until President Clinton terminated it in May 2000. However, as the Europeans quickly point out, the United States still maintains the capability to degrade civilian GPS accuracy immediately upon direction.[4] Thus, many Europeans fear (or claim to fear) that the "military character of GPS means there is always a risk of civil users being cut off in the event of a crisis."[5] Although technically possible, such an occurrence is remote because the United States has never degraded (beyond established selective-availability policy levels) or removed GPS signals during wars or crises. For instance, the United States neither increased degradation beyond the standard 100 meters during operations in Bosnia and Kosovo nor reactivated selective availability during post-9/11 operations in Afghanistan and Iraq.[6] Nevertheless, even with selective availability turned off, GPS accuracy does not meet requirements for all civil applications.

Although GPS civilian accuracy suffices for recreational and many other functions, accuracy of 10–20 meters provided by a single signal does not meet requirements for sole-means navigation in safety-critical applications such as entering a seaport or landing aircraft under poor weather conditions.[7] Moreover, "current space-based radio navigation systems do not provide adequate performance to meet European multi-modal and multi-sector needs."[8] For that, Europe's use of GPS would require heavy investments in differential technology in which line-of-sight limitations constrain use to localized areas (without additional investments in Wide Area Augmentation Systems [WAAS]).[9] Alternatively, the proposed Galileo system will transmit multiple civilian signals, providing a global accuracy of approximately four meters.[10] Additionally (as discussed in chap. 2), GPS accuracy degrades at high latitudes and in urban settings. The Center for Transport Studies at Imperial College, London, tested the urban-canyon problem in the English capital and discovered that GPS five-meter accuracy was available only 17 percent of the time.[11] Galileo's orbit design hopes to improve accuracy for Nordic users, but it is unlikely to overcome the urban-canyon occultation problem on its own. The largest improvement in accuracy for any environment would come not from a new system, but from the combination of GPS and Galileo systems. More satellites in orbit will increase the "probability of having a clear view to sufficient satellites for a robust positioning solution."[12] However, while "mediocre and varying position accuracy" is an issue, Europeans are more concerned with the availability and reliability of GPS signals.[13]

The GPS civilian service, the SPS, is not guaranteed worldwide at all times. In fact, it is not guaranteed at all. Nor does it quickly inform users of substandard performance. While GPS is generally reliable and system failures are rare, outages have caused discontinuous service in the past. In 2000, for example, GPS satellite malfunctions deprived the areas of Oklahoma, Kansas, and Nebraska of navigation signals for 18 minutes.[14] In other cases, a Canadian research body reported that one aircraft was affected by an unannounced signal interruption greater than 80 minutes, and Icelandic aviation authorities noted that several transatlantic flights in their control zone were similarly disturbed.[15] Likewise, the deactivation of a satellite for maintenance shut down a series of automated bank-teller machines as well as a communications network that relied on GPS for synchronization.[16]

Consequently, if satellite navigation is a keystone of transportation infrastructure, even minor service discontinuities can have severe consequences for safety. For this reason the European transport industry is a primary driving force behind Galileo.[17] Not surprisingly, Europeans have identified satellite navigation as an essential tool for implementing the European Transport Policy.[18] Extant terrestrial navigation aids vary widely in number and technology across Europe, and each transport community uses different types of systems without a coordinated policy at the European level.[19] Europe plans to employ a GNSS to standardize and "harmonize future transport guidance systems."[20] Numerous safety-critical applications will depend on the GNSS. Accordingly, EU directives for safe transport of people and goods require a GNSS to provide service guarantees with liability commitments.[21] Given that

GPS does not provide such guarantees, Europe is seeking a separate GNSS.

The last European concern regarding GPS performance is vulnerability. As confirmed in the US DOT's Volpe Report, GPS is susceptible to intentional (jamming) and unintentional interference. As previously noted, GPS is easily jammed because it employs extremely low-power signals. For the same reason it is also susceptible to unintentional disruption from mobile-phone satellite systems, television broadcasts, and natural phenomena such as ionospheric interference and solar flares.[22] This situation was illustrated when interference affecting GPS receivers and differential stations in and around the harbor of Monterey, California, was traced to a fishing boat and two other sources using active television antennas emitting in the L1 frequency range.[23] In view of these vulnerabilities, the Volpe Report states that GPS cannot serve as a sole source of PNT services for critical applications and that backup systems are vital for all GPS applications involving the potential for major economic or environmental impacts or SoL situations.[24]

The United States has not sat idle in the face of these performance drawbacks. The ongoing GPS modernization plan will eventually address all of these shortcomings. Next-generation satellites—the Block IIR-M—will add a second civilian signal on L2 (referred to as L2C), greatly increasing civilian accuracy. Block IIR-M will also debut the military M-code on L1 and L2, "which will provide improved signal-processing techniques for enhanced jamming resistance."[25] Block IIF satellites will add a third civilian signal, L5, featuring more power and a new coding scheme. The signal will increase availability and civilian accuracy, potentially to the centimeter level.[26] The United States plans to begin launching Block IIF satellites in 2006.[27] GPS III satellites will have 100 to 300 times the transmission power of the current constellation to mitigate jamming and interference.[28] They will also debut high-gain antennas to generate directional spot beams several hundred kilometers in diameter.[29] Spot beams will allow the US military to focus more power in particular regions to resist jamming.[30] Lastly and perhaps most importantly, GPS III will provide integrity monitoring, increasing reliability by quickly informing users of performance degradation.[31] Unfortunately, GPS III will not begin launching until 2012.[32] In due course, GPS will address all these stated performance concerns regarding accuracy, reliability, and vulnerability. However, Europe and the rest of the world will not wait—they plan to act independently.

Independence

An increasingly integrated Europe has progressively sought to "acquire power and project geopolitical ambition," especially since the end of the Cold War.[33] One of the EU's chief goals is to create "a superpower on the European continent that stands equal to the United States."[34] Naturally, this ambition extends to space. As early as 1991, the EC hinted at the potential development of an independent navigation system to reduce European dependence on US space-defense systems.[35] Unsurprisingly, the ESA claims that "European independence is the chief reason" for building Galileo.[36] Indeed, Galileo strengthens Europe's bid for political, security, and technological independence from the United States.

Political Independence

Originally, Europe did not plan to build its own satellite-navigation system. It had hoped to participate in an internationally developed global system similar to GPS for civilian use "under the aegis of the United Nations."[37] However, the United States concluded that this idea was not in its best interest. It would not cede control of GPS and was not interested in running an additional system, effectively stalling the development of a global system under civilian control and, ultimately, planting the seeds for Galileo.[38] In the meantime, GPS flourished.

To the extent that it affects sovereignty, the impact of GPS on modern European society is pervasive. Europe plans to employ a GNSS to aid the implementation of a broad set of policies that includes regulating agriculture, fisheries, and transportation services.[39] For example, Europeans forecast high growth rates for inland transport that will double loading by 2020.[40] This increase is expected to overwhelm the existing infrastructure.[41] The European GNSS is expected to provide a sustainable transport policy by monitoring traffic flows, preventing congestion, and enabling automatic toll payments without stopping for tollgates.[42] Thus, GNSS technology will become increasingly embedded in European domestic policy, forming "the basis of important commercial applications and government-supported infrastructures."[43] In the absence of Galileo, these basic governmental decisions and policies would depend upon US policy for GPS, which is subject to change without reference to European requirements. Without Galileo, European critical infrastructure will rely on a system owned and operated by a foreign power's military. This potential condition conjures up fearful images that "Europe can be held to ransom on all issues related to its use of GPS and might be obliged to pay governmental levies to the US in the future."[44] Additionally, under various international conventions, EU member states have public obligations to provide safe navigation and certain other public services, such as SAR.[45] These European states have laws requiring them to control and regulate navigation aids used in SoL applications, but Europe cannot control or regulate GPS.[46] Ultimately, the European Parliament (EP) concluded that "the sovereignty and safety of Europe will be in serious danger if the European navigation systems are removed from European control."[47]

In addition to sovereignty, a powerful political motive spurring the pursuit of Galileo is prestige, influencing the development of policies that allocate status and acknowledge achievement.[48] In 1957 the launch of Sputnik instigated an international space race—essentially a superiority contest between the United States and USSR to garner influence over the rest of the world. Many believed that "emerging Third World nations would follow the country considered most technically advanced, for development and political reasons."[49] However, "since the end of the Cold War, the stakes in the space race have shifted from prestige . . . towards market shares and dominance for applications."[50] Today space activities are judged by what they provide to society in the broad sense, especially economically.[51]

Nonetheless, Europeans still regard space as an area for significant technological innovation that can elevate international standing.[52] Similar to the impacts of Ariane and Airbus, Europeans believe Galileo will enhance the international influence of the EU.[53] Galileo will help build leading-edge technology and a strong economy, prime assets regarding Europe's influence and attractiveness in the world.[54] According to an EU government advisor, "There was a prestige aspect involved in the whole development [of Galileo], that Europe wanted to show that they could indeed do something."[55] Epitomizing the influence of vying for status, French president Jacques Chirac warned that Europe's failure to develop Galileo "would inevitably lead to [Europe] becoming . . . vassals" of the United States.[56]

Additionally, anti-American sentiment on the part of European elites may be augmenting the pursuit of Galileo.[57] Inflamed by America's unilateralist expansion of the war against terrorism to Iraq and the explicitly preemptive rhetoric contained in the 2002 *National Security Strategy of the United States*, some European leaders have grown disgruntled "with America's might and its overbearing ways."[58] Consequently, Europe may symbolically view Galileo as "a means to struggle against American hegemony."[59] Regardless of the political motive, would the United States place its policies for national security and economic development on critical infrastructure owned by Europe?

Security Independence

The European security perspective has changed over the years, and Galileo will play an important role in the future defense of Europe. Europe has depended on the United States for security since the end of World War II. NATO was designed to keep the "Americans in, the Russians out, and the Germans down."[60] However, the end of the Cold War changed the geopolitical landscape. In Europe today, the Americans are superfluous, the Russians are irrelevant, and the Germans are integrated. European security has faded as an American priority in the absence of the USSR, as illustrated both by Congress's insistence that Europe bear a greater defense burden and by America's reluctance to prosecute the war in Kosovo.[61] Certainly, the post-9/11 environment refocused American priorities on homeland defense and the war on terrorism.

While originally slow to respond to these changes, Europe has redoubled its efforts to build a common defense policy in the wake of the Kosovo campaign, an indication that "the Europeans are scared . . . that America will not show up the next time war breaks out somewhere near Europe's periphery."[62] Indeed, by 2000 the EU established the position of High Representative for Common Foreign and Security Policy and committed itself to fielding a rapid-reaction force of 60,000 troops deployable for at least a year to conduct peace operations.[63] After the Kosovo war, "several European governments agreed that an autonomous satellite navigation capability must serve as the basis for Europe's security and defense policy."[64] Thus, Europe is taking steps to end its security dependence, and Galileo will figure prominently in the endeavor. However, supporting military operations goes against a founding principle of Galileo.

Europe insists Galileo is designed "specifically for civilian purposes," as compared to GPS, which was "designed during the Cold War for military purposes."[65] By contrasting Galileo's peaceful orientation with GPS's military roots Europe implies that Galileo is the best choice for civilians, since meeting civilian needs is not the Pentagon's top priority.[66] The European Parliament expressly stated that "[Galileo] is being developed by civilian organizations . . . and run under civilian control. . . . [Galileo] is not designed for specifically military purposes."[67] Some EU states were quite emphatic that Galileo remain strictly civilian in nature.[68] In particular, the British insisted that all public statements about Galileo stress civilian applications.[69] Moreover, they did not see the need for the PRS, Galileo's encrypted service with potential military value.[70] Essentially, the Galileo doves felt that "using Galileo for military purposes would jeopardize business investments."[71] In contrast, France has quietly championed Galileo for military purposes and has previously threatened to withdraw support for it without the PRS.[72] Historically, France led European efforts for autonomy, marked by its independent procurement of nuclear weapons in the face of American nonproliferation policy, its departure from NATO's military command structure, and its push for the Airbus and Ariane programs. In fact, France believed that "Europe as a whole was destined to eventually move away from military, economic, and technological dependence on the US."[73] As the giant of the European space industry, France used its influence to help move the EU's position on Galileo from a civil system under civil control, to a civil system usable by the military, to a system critical for European defense.[74] In the end, the Directorate-General for Energy and Transport (DG-TREN) acknowledged, "Although designed primarily for civilian applications, Galileo will also give the EU a military capability."[75]

GNSS technology has proven essential to military capabilities, as illustrated by the impact of GPS on US military operations. During Operation Desert Storm, GPS guided coalition forces through the featureless desert, from the tanks involved in the famous left-hook maneuver to the Apache and Pave Low helicopters that provided the opening salvos of the war.[76] During Operation Allied Force (OAF), GPS-guided munitions became a requirement, enabling precision all-weather bombing of Kosovo and Belgrade.[77] Without GPS, pilots facing 50 percent cloud cover more than 70 percent of the time would have to wait for clearer skies, creating sanctuaries and operational lulls.[78] GPS-guided munitions proved so effective that US reliance on them increased from 3 percent of all bombs during OAF to roughly 60 percent in Operation Iraqi Freedom (OIF).[79]

In addition to precision navigation and munitions, GPS also provided enhanced battlefield awareness. During OIF the US Army and special operations forces utilized Blue Force Tracking (BFT) systems that integrated GPS with other space technology to provide theaterwide situational awareness. BFT significantly reduced fratricide in OIF, during which only one soldier was killed by friendly direct ground fire, as compared to 35 deaths in Desert Storm.[80] Also, BFT proved essential to the famous "race to Baghdad," greatly accelerating the tempo of combat through battlespace awareness and real-time information.[81] Without it the maneuver would have lost cohesion because the vehicles moved so fast they outran the range of their radios.[82] Consequently, space activities such as GPS have become indispensable to US national security.[83]

Europeans clearly understand the value of a GNSS to security and view Galileo as a means to "hedge against the perceived risk that the US Department of Defense would deliberately degrade or jam a signal increasingly vital to European interests."[84] The first line of defense for Europe, NATO has sanctioned GPS as its primary navigational aid.[85] But not all EU states are members of NATO. There may come a time when the EU acts on its own, outside of NATO and possibly against US interests. According to the DG-TREN, "If the EU finds it necessary to undertake a security mission that the US does not consider to be in its interest, [Europe] will be impotent unless it has the satellite navigation technology that is now indispensable."[86] In other words, "Galileo will underpin the common European defense policy."[87]

Technological Independence

Since the 1950s "the drive for advanced technology . . . [has been] and remains a key motivation for European activities in space" and serves as a primary incentive fueling the development of Galileo.[88] While early European space efforts were tied to cooperation with NASA, Europe sought more than just a cooperative role in space; "the Europeans [wanted] to master the technology of space activity."[89] Indeed, "the most fundamental reason for cooperation . . . was to help European industry develop its know-how and potential."[90] This was especially important because Europeans perceived that a technology gap, created by heavy US spending on research and development, divided America and its allies and provided American firms leverage in European markets.[91] In the early 1960s the fear of falling behind led the Europeans to establish "their own space research organizations . . . precisely to promote European competitive independence from the United States in advanced technology."[92] One organization explicitly advocated that Europe should forgo importing US space technology so that Europeans might acquire experience in research and development.[93] Today Europe looks to Galileo to serve this exact purpose. Without access to technological developments in the satellite-navigation sector, the EU's industrial capacity to compete in the US-dominated market would be seriously constrained—a lesson learned from previous experience.[94]

Galileo is not the first European venture designed to overcome US technological dominance in space. In the 1960s the United States enjoyed an almost total monopoly in communication satellites and sought to control international developments in the field through the creation of the International

Telecommunication Satellite Organization (INTELSAT).[95] Predictably, the provisions of INTELSAT nourished American hegemony and heightened the technology gap, galvanizing the Europeans to move away from scientific projects to pursue their own "technologically relevant, commercially viable endeavors" such as the Symphonie satellite communication program.[96] However, as the Europeans shifted away from scientific spacecraft with hopes to achieve some measure of technological independence, they ran headlong into additional obstacles imposed by a near-monopoly of American launch systems. In 1972 the United States prohibited the export of space-launch technology "in support of its own launch providers . . . creating a clear state of dependence by other countries."[97] Previous US policy pledged to launch European spacecraft only of a scientific or experimental nature.[98] As a result, the United States refused to launch Symphonie until "assured that . . . [it] would not compete with Intelsat for commercial traffic."[99] The United States eventually launched the satellite but not until the Europeans begrudgingly acquiesced to the required concessions.[100] Thus, to escape their technological dependence on the United States and America's unwillingness to guarantee space-launch services, the Europeans independently pursued development of the Ariane launch booster.[101] Fast-forwarding to the present, US dominance in satellite-navigation technology once again threatens Europe with technological dependence and has spurred the pursuit of Galileo.

According to Alenia Spazio, a major Italian space company, as recently as three years ago Europe had no industry for satellite navigation.[102] The market was and remains dominated by GPS and US industries. Accordingly, Carl Bildt, a former Swedish prime minister, contends that "the most important reason for Europe to develop Galileo is to maintain Europe's high-tech industrial base."[103] Europeans believe that developing Galileo will help build technical skills and knowledge on a learn-as-you-go basis as engineers meet the challenges of satellite and ground-systems design, manufacturing, and certification.[104] For example, European engineers will enhance their competency in "space-qualified clocks and volume parts procurement for multi-satellite constellations," skills transferable to various other space applications.[105] Still fearing the technology gap, Europe feels it "cannot allow itself to lag behind in the future development of technological capacities and the management of related technologies."[106]

Most significantly, technological independence enables the ability to influence or set system standards. "Whoever defines the requirements, develops the specifications and sets standards for [Galileo] satellite signals and equipment will have tremendous commercial leverage."[107] For example, in the 1980s, Europe lagged considerably behind the United States in digital cellular telephony.[108] In response, the EU "supported the launch of the Global System for Mobile Communications, which ultimately set the new digital standard," enabling European companies like Nokia and Ericsson to lead the industry.[109] Certainly, the Europeans feel that their "absence from the definition of the [GPS] space segment" significantly hurts their position "in the rapidly expanding markets for user equipment and value-added services."[110] Furthermore, "experience shows that only those nations and industries with a decisive influence on the system infrastructure will remain competitive in the market."[111]

Economic Opportunity

The third pillar of Europe's three-tiered European Space Strategy is to reap the benefits of space for markets and society.[112] Germany's federal minister of education and research underscored this objective by asserting that "the most important goal of the European Space Strategy is the consistent use of space technologies to seize market opportunities."[113] In addition to concerns about GPS performance and European independence, economic opportunity emerges as a major incentive to develop Galileo.

The market for GNSS civil applications is immense and growing rapidly. In 2002 "commercial services based on free access to GPS [had] revenues estimated at around $12 billion," and the global market for services and receivers is forecasted to approach Euro 40 billion by 2005.[114] By 2020 Europeans estimate "that over 65 percent of the population of Europe will rely on GNSS while going about their business and daily lives," driven largely by an anticipated surge in personal GNSS use in vehicles and services integrated with mobile phones.[115] Currently, however, Europe's share of the satellite-navigation-terminal market is minuscule, amounting "to 15% in Europe, and only 5% worldwide."[116] Moreover, a study in 2000 that examined European competitiveness in the GNSS industry reveals that "Europe had few suppliers of GPS chipsets/receivers for high-end applications as well as low-end mass market products."[117] In fact, during that time over 80 percent of the GPS receivers were designed and manufactured in the United States.[118] With Galileo, Europeans believe they will gain a foothold in the market, much like they did with Ariane.[119] Today, Ariane commands nearly 50 percent of the commercial-satellite-launch market.[120]

If Europe can establish a foothold, sales of Galileo receivers are expected to increase from "[Euro] 100 million in 2010 to some 875 million by 2020, representing market penetration rising from 13 percent to 52 percent."[121] Increasing the European share of the satellite-navigation market via Galileo would drive the creation of jobs. Studies performed for the EU vary in their predictions, ranging from 100,000 jobs by 2020 to 146,000 by 2025.[122] In addition to driving up market share and creating jobs, Galileo is forecast to begin turning a profit by 2011 through royalties and service charges.[123] Given these predictions, a substantial market exists in the future, and, according to the EU's vice president, "The challenge is to ensure that Europe can take a fair share of this global market."[124] For an economically competitive Europe, Galileo provides a window of opportunity—but the window will not stay open long.

According to an independent study by Pricewaterhouse-Coopers, Galileo must commence service by 2008 because "the market will be in a rapid growth phase by then, and GPS III . . . is expected to commence operations one or two years thereafter. Galileo will only become established if it is in the market with enough time to gain acceptance in the launch of new equipment and services."[125] If launched as planned (an assumption most economic studies rely upon), Galileo will provide improved accuracy and integrity monitoring several years before GPS can introduce comparable services. As discussed previously, GPS III will provide virtually the same services as Galileo for free "and would thus close the window of opportunity for Europe to set the global standard in satellite navigation."[126] Because the United States does not plan to launch GPS III before 2012, however, Galileo has a four- to five-year window to establish itself with a superior product. Considering that most space programs encounter delays during development, any setback for Galileo could nullify its competitive advantage.[127] Accordingly, the EU fears that if it postpones Galileo, the market will adopt GPS as the standard, relegating Europe to a supporting role.[128] Hence, Galileo is a classic case of the need to "git thar fustest with the mostest" in order "to combat the USA's current monopoly."[129]

In general, monopolies stifle "technological innovation and economic progress," and so the US monopoly on satellite navigation services provides additional economic incentive to launch Galileo.[130] Europe believes the "absence of competition means that optimal service cannot be provided for private users, neither can free reception be guaranteed in the long term."[131] It looks to Galileo to remedy the US monopoly, just as Airbus challenged the Boeing monopoly and "brought airlines, passengers, and crews the benefits of real competition."[132] The same may be said for Ariane and the commercial launch industry. To at least some degree, it can be argued that the rise of Galileo is America's own fault.

In 1995 Irving Lachow warned that "international acceptance of GPS is important from an economic and commercial standpoint because the lack of it could lead to competing satellite navigation systems."[133] With no competition to drive improvements for civilian users, the Pentagon focused on refining GPS military applications and did not satisfy major concerns of the civilian-user community, especially foreign governments.[134] A lack of competition allows the United States to follow a launch-on-need policy, whereby it replaces GPS satellites only when they fail.[135] GPS satellites have routinely outlived their design life, however, and this has had the unintended result of preserving old technology in orbit and, for economizing, holding improvements hostage to the failure rate.[136] One does not wait until the family computer ceases to operate before purchasing an upgraded model to exploit faster processing or other new features. Nevertheless, it is likely that the status of the on-orbit Block IIR constellation will influence the launch of Block IIF satellites, which would delay the debut of L5, the third civilian signal.

Additionally, Lachow warned that "the technologies required to develop and deploy a satellite navigation system are no longer state-of-the-art," falling well within the capabilities of private companies.[137] The biggest obstacle preventing other players from entering the satellite-navigation market is investment capital, which the Europeans have overcome through collective action.[138] Consequently, one cannot fault Europe for pursuing Galileo because "societies which fail to maintain competitiveness run the danger of economic stagnation and eventual decline."[139]

In short, while individual motives for Galileo may vary depending on the nation, collectively the EU is pursuing it for improved performance, independence from the United States, and economic opportunity. Specifically, Europe believes Galileo will outperform GPS and is inherently more accurate, more reliable, and less vulnerable by design. With Galileo, Europe can secure a degree of political, security, and technological independence from the United States. Finally, Galileo offers Europe an economic window of opportunity to seize the satellite-navigation market from the United States and to set a new global standard. While Galileo should have a considerably positive impact on Europe, the ramifications of Galileo's development and implementation may have an equally significant impact on the United States.

Notes

1. The White House, Office of Science and Technology Policy, "US Global Positioning System."
2. Dinerman, "GPS and Galileo," 10.
3. European Commission, "European Dependence," technical note, 2.
4. Ibid.
5. Wilson, *Galileo: The European Programme*, 7.
6. US Department of State, "US Global Positioning System." The United States has discontinued selective availability because it has developed methods to deny enemy use of GPS signals in localized areas. See *Public Papers: Clinton*, 803.
7. Wilson, *Galileo: The European Programme*, 8.
8. de Palacio, "Importance of Galileo for Europe."
9. Lembke, *Competition for Technological Leadership*, 69.
10. European Commission, *Galileo: Mission High Level Definition*, 14.
11. Byrne, "What's So Wrong with GPS?" 32.

12. European Commission, *European Dependence*, 27.
13. Wilson, *Galileo: The European Programme*, 7.
14. European Commission, *European Dependence*, 1.
15. Wilson, *Galileo: The European Programme*, 7.
16. Dornheim, "GPS Improvements Set," 56.
17. European Parliament, *Report on the Commission Communication to the European Parliament*, 12.
18. European Commission, *European Dependence*, 3.
19. European Commission, *Galileo*, 8.
20. de Palacio, "Importance of Galileo for Europe."
21. European Commission, *European Dependence*, 3.
22. US Department of Transportation, "Vulnerability of the Transportation Infrastructure," 25–27.
23. Dornheim, "GPS Improvements Set," 56.
24. US Department of Transportation, "Vulnerability of the Transportation Infrastructure," 6.
25. Sirak, "USA Sets Sights on GPS," 30.
26. Dornheim, "GPS Improvements Set," 56; and Kirk Lewis (senior analyst, Institute for Defense Analyses [IDA], Alexandria, VA), interview by the author, 9 December 2003.
27. Hewish, "What Is Happening with GPS?" 54.
28. Sirak, "USA Sets Sights on GPS," 30.
29. Hewish, "What Is Happening with GPS?" 54.
30. Sirak, "USA Sets Sights on GPS," 30.
31. Dornheim, "GPS Improvements Set," 56.
32. Ibid.
33. Kupchan, *End of the American Era*, 151.
34. Reid, "EU Summit Ends," *Washington Post*, 17 March 2002.
35. European Commission, *European Community*, 27.
36. European Space Agency, "Why Europe Needs Galileo."
37. Lembke, *Competition for Technological Leadership*, 59.
38. Ibid.
39. European Commission, *European Dependence*, 3.
40. Lembke, "EU Critical Infrastructure," 108.
41. European Parliament, *Report on the Commission Communication to the European Parliament*, 12–13.
42. Ibid, 13.
43. Bulmahn, "Europe's Ambitions in Space."
44. European Commission, *European Dependence*, 6.
45. European Commission, *Galileo: Involving Europe*, 2.
46. Pace et al., *Global Positioning System*, 39.
47. European Parliament, *Report on the Commission Communication to the European Parliament*, 15.
48. Handberg and Johnson-Freese, *Prestige Trap*, 4.
49. Ibid., 20.
50. European Commission, "Towards a Coherent European Approach," 5.

51. Handberg and Johnson-Freese, *Prestige Trap*, 2.
52. European Parliament, *Report on the Commission Communication to the European Parliament*, 9.
53. Commission of the European Communities, *Commission Communication to the European Parliament*, 8.
54. European Commission, "Galileo: A Decision," 6.
55. Lembke, *Competition for Technological Leadership*, 99.
56. Braunschvig et al., "Space Diplomacy," 160.
57. Dinerman, "GPS and Galileo," 10.
58. Kupchan, *End of the American Era*, 154.
59. Ibid. Kupchan quotes French president Jacques Chirac.
60. Robertson, referring to Lord Ismay's famous quote in "NATO in the 21st Century."
61. Kupchan, *End of the American Era*, 152.
62. Ibid.
63. Ibid., 149.
64. Braunschvig et al., "Space Diplomacy," 159.
65. Wilson, *Galileo: The European Programme*, 5, 7.
66. National Academy of Public Administration, *Global Positioning System: Charting the Future*, 42.
67. European Parliament, *Report on the Commission Communication to the European Parliament*, 14.
68. Bell, "GPS and Galileo."
69. "Europe: Eppur Si Muove," 53.
70. Divis, "Galileo Breaks Free."
71. Lembke, "Politics of Galileo," 10.
72. Karner, interview (see chap. 2, n. 3).
73. Handberg and Johnson-Freese, *Prestige Trap*, 171.
74. Karner, interview.
75. European Commission, "Galileo: A Decision," 6.
76. Mackenzie, "Apache Attack."
77. Air Force Doctrine Document (AFDD) 2-2, *Space Operations*, 51.
78. Ibid.
79. Braunschvig et al., "Space Diplomacy," 158.
80. Dunn, "Blue Force Tracking," 11.
81. Robinson, "Who Goes There?"; and Peartree et al., "Information Superiority," 120.
82. Robinson, "Who Goes There?"
83. Condoleeza Rice, quoted by Butler in "Rice Wants President to Initiate."
84. Braunschvig et al., "Space Diplomacy," 159.
85. Bell, "GPS and Galileo."
86. European Commission, "Galileo: A Decision," 6.
87. Ibid.
88. Handberg and Johnson-Freese, *Prestige Trap*, 21.
89. Frutkin, *International Cooperation in Space*, 133.

90. Handberg and Johnson-Freese, *Prestige Trap*, 172. Handberg quotes A. Dattner, *Reflections on Europe in Space: The First Two Decades and Beyond* (Noordwijk, Netherlands: ESA Scientific and Technical Publications Branch, ESTEC, 1982).

91. Sebesta, "US-European Relations," 139.
92. McDougall, . . . *the Heavens and the Earth*, 208.
93. Ibid., 426.
94. European Commission, *Galileo: Involving Europe*, iv.
95. Sebesta, "US-European Relations," 141.
96. Ibid., 140.
97. Handberg and Johnson-Freese, *Prestige Trap*, 170.
98. Sebesta, "US-European Relations," 150.
99. McLucas, *Space Commerce*, 54.
100. Sebesta, "US-European Relations," 155.
101. Handberg and Johnson-Freese, *Prestige Trap*, 170.
102. Lembke, *Competition for Technological Leadership*, 123.
103. Grant and Keohane, "Europe Needs More Space," R14.
104. European Commission, "European Dependence," technical note, 7.
105. Ibid.
106. European Parliament, *Report on the Commission Communication to the European Parliament*, 6.
107. Lembke, "Politics of Galileo," 22.
108. Braunschvig et al., "Space Diplomacy," 162.
109. Ibid.
110. European Commission, "Towards a Coherent European Approach," 17.
111. European Commission, *European Union and Space*, 23.
112. European Commission, Communication from the Commission to the Council, *Europe and Space*, 3.
113. Bulmahn, "Europe's Ambitions in Space."
114. "Business: Navigating the Future," 92; and European Parliament, *Report on the Commission Communication to the European Parliament* for the 2005 estimate.
115. European Commission, *European Dependence*, 14–15.
116. European Parliament, *Report on the Commission Communication to the European Parliament*, 11.
117. European Commission, *European Dependence*, 17. The document refers to the "Structural Analysis of the European Satellite Application Segment" study from Technomar GmbH, October 2000.
118. Lembke, *Competition for High Technology*, 123.
119. Commission of the European Communities, *Commission Communication to the European Parliament*, 29.
120. European Parliament, *Report on the Commission Communication to the European Parliament*, 9.
121. PricewaterhouseCoopers, *Inception Study*, 4.

122. See de Palacio, "Importance of Galileo for Europe," for 100,000 jobs; and Lembke, "EU Critical Infrastructure," 109 for 146,000 jobs. The numbers include employment in design, manufacturing, operations, sales, and services sectors.
123. PricewaterhouseCoopers, *Inception Study*, 3, 7.
124. de Palacio, "Importance of Galileo for Europe."
125. PricewaterhouseCoopers, *Inception Study*, 4.
126. Braunschvig et al., "Space Diplomacy," 160.
127. "Europe: Eppur Si Muove," 53.
128. European Commission, *Galileo: Involving Europe*, 44, 54.
129. General Nathan Bedford Forrest Historical Society, "Quotes by General Forrest"; and European Parliament, *Report on the Commission Communication*.
130. Gibbons, "World of Difference."
131. European Parliament, *Report on the Commission Communication to the European Parliament*, 10.
132. "Airbus Today."
133. Lachow, "GPS Dilemma," 141.
134. Ibid.
135. Divis, "GPS III," 10.
136. Ibid.
137. Lachow, "GPS Dilemma," 141.
138. MacDonald, "Econosats," 44–54.
139. Handberg and Johnson-Freese, *Prestige Trap*, 1.

Chapter 4

Implications and Recommendations

Failure to master space means being second best in every aspect.

—Pres. Lyndon B. Johnson

Today, through lack of focus and funding, the United States stands to lose not only its primacy but even its capability in satellite navigation if it does not rise to the occasion.

—David Braunschvig

The European public and private sectors, driven by the motives outlined in the previous chapter, have provided the necessary financial and political backing for Galileo to proceed to the development phase. Additionally, Galileo has attracted interest and investment from many non-European nations, including the People's Republic of China. With this groundswell of international support, Galileo is fast becoming reality. Assuming Europe implements Galileo as planned, the implications for US space policy are significant, and its response will be carefully monitored around the world. In this final chapter, I examine the national security and economic concerns generated by the emergence of Galileo, review US policy towards Galileo, and provide recommendations for the future.

Implications

The primary goals of the US national policy for GPS are to strengthen and maintain US national security and to support and enhance US economic competitiveness and productivity.[1] As currently designed and promoted, the proposed Galileo system directly challenges these US national interests. Since GPS is now integrated into virtually every facet of US military operations, anything that potentially interferes with GPS threatens national

IMPLICATIONS AND RECOMMENDATIONS

security.[2] As a result, the advent of Galileo sparks security concerns centering on space control and superiority.

The Air Force defines space superiority as the "degree of control necessary to employ, maneuver, and engage space forces while denying the same capability to an adversary."[3] This includes protecting the benefits of space support to the friendly war fighter as well as denying these same benefits to the enemy. One tool of space superiority is navigation warfare. The concept of NAVWAR, initiated in 1996 by the United States, protects US as well as allied use of GPS during conflicts, prevents the enemy from exploiting GPS, and preserves normal signal availability outside the theater of operations for global civilian users.[4] Essentially, the United States plans to jam GPS (or any other satellite-navigation system) to deny the enemy's access in a localized area. Currently, civilian and military signals share the same GPS frequency on L1, and military receivers utilize the civilian signal to acquire the more complex military signal.[5] Thus, jamming L1 would effectively reduce military accuracy.[6] For this reason, in order to optimize NAVWAR, future GPS satellites (beginning with Block IIR-M) will transmit military signals (M-code) spectrally separated from civilian signals, eliminating the potential for signal fratricide and enabling more efficient jamming. However, at the 2000 World Radiocommunications Conference, the International Telecommunications Union authorized Galileo to transmit its PRS and OS signals in the same frequency range as the GPS M-code. Furthermore, Europeans planned to transmit the PRS signal using the same modulation scheme as the GPS M-code, thereby directly overlaying Galileo's PRS signal on top of the GPS M-code signal. In addition to interfering with GPS signals, any attempt by the United States to jam the PRS would also jam the M-code, effectively nullifying NAVWAR.

As a result, US Deputy Defense Secretary Paul Wolfowitz emphasized US concerns in a letter to NATO defense ministers, stating that "the addition of any Galileo services in the same spectrum . . . will significantly complicate our ability to ensure availability of critical GPS services in time of crisis or conflict and at the same time assure adversary forces are denied similar capabilities."[7] Subsequently, George Bell, NATO's assistant secretary general for defence support, reiterated these security concerns, stating that the signal-overlay condition produced "a negative impact on NATO's military effectiveness in the area of operations, potentially risking fratricide on friendly forces and civil populations."[8] Not surprisingly, the EU viewed things differently.

The Europeans insisted on pursuing the proposed PRS signal specification because it "offers the best performance in peacetime, particularly in terms of resistance and robustness."[9] Arguments for optimum signal robustness aside, by overlaying the PRS on the M-code, Europeans could force the United States to include them in jamming decisions.[10] In fact, the EU asserted that "a political agreement on the cooperation necessary between the two radionavigation systems is required in preparation for a crisis."[11] With this pronouncement, it is clear that the EU is positioning itself to be consulted before the United States implements NAVWAR and jams Galileo.[12] In November 2003, US-EU negotiations appeared to resolve the PRS overlay problem in principle; however, the final terms have yet to be agreed upon.[13] Furthermore, some believe that the EU's concession could be temporary and that the EU, driven by France, intends to change the frequency once it has demonstrated its ability to manage an encrypted signal.[14] Despite negotiations, the French remain committed to a direct PRS overlay of the M-code, in part because they plan to incorporate Galileo into weapons manufactured for export, and because an unjammable signal would undoubtedly boost sales in the arms market.[15] Indeed, the French firm Thales already offered a presentation on PRS-based military-equipment markets at the Institute of Navigation's annual conference in 2002.[16]

Besides posturing for a joint US-EU decision process for denial of service, the EU believes that the need to jam the PRS is negligible because the signal will be encrypted and restricted to authorized users.[17] In response, NATO highlighted concerns regarding the integrity of the PRS encryption regime, fearing that PRS signals could be compromised and exploited by an adversary.[18] Likewise, the United States fears that rogue states, terrorists, or even states acting against US interests could use Galileo to their advantage.[19] Although the EU avers that PRS access requires a special cryptographic key that will be strictly controlled by key-management systems approved by EU member states, the EU has not yet addressed the specifics of the crypto-security regime or who exactly will have access.[20] According to Bell, "Detailed discussion on the crucial issues . . . related to the control and possible proliferation of user equipment, the robustness of associated cryptography and distribution and control procedures for the keys have [sic] not been initiated or authorized between NATO and the EC."[21]

For the moment, the EU believes that the United States should trust Europe to secure its PRS signal. Hoping to convince the United States of the encrypted service's viability, the EU stated that "some EU Member States have the know-how to design and implement effective government encryption. The resulting technology could be made available to the European authorities controlling the Galileo PRS signal."[22] In this manner, Galileo would mitigate fears of a PRS compromise and "ensure signal denial to hostile nations where necessary."[23] The EU asserts that European nations within NATO have trusted the US security mechanisms for GPS ever since a 1993 memorandum of understanding between the US DOD and NATO provided them access to the PPS.[24] Consequently, "the EU would like the US to show the same trust regarding its capability to implement a secure Galileo system."[25]

Also worrisome for the United States is that in the process of seeking additional funding and support for Galileo, the EU has welcomed non-European investment and participation in Galileo's development, further complicating the issue of controlled access. China, India, Israel, Canada, and South Korea have expressed interest in assisting the EU, with some degree of access and influence in return.[26] In particular, China's prime minister has "expressed his country's interest in being fully involved in the Galileo programme financially, technically and politically."[27] In fact "China's support . . . can facilitate EU's negotiation with the US . . . on cooperation" by providing the EU with additional diplomatic leverage.[28] An EC minister confidently stated: "We expect Chinese support to our positions on frequencies and international standardization activities."[29] The issue is more than merely speculative. China recently pledged Euro 200 million ($236 million) for Galileo and is primarily interested in investing in the PRS.[30] The prospect

IMPLICATIONS AND RECOMMENDATIONS

of China with access to an encrypted and potentially unjammable navigation service raises concerns among US military and foreign policy officials.[31] Moreover, through continued investment and support, China could possibly buy a seat on the Galileo security committee that controls access to the PRS—a committee currently conceived to decide issues via unanimous vote.[32] Although the EU insists that China will not have access to the PRS or any other security aspects of the system, the extent of China's participation will be determined in subsequent agreements and will likely be influenced by "how [China's] initial investment takes shape."[33] Beyond China's direct investment and support, access to its 100 million mobile users provides a huge market for Galileo.[34]

Along with national security issues, Galileo challenges US economic competitiveness, with implications for fair competition and assured access to the global satellite-navigation market. The US national policy for GPS encourages worldwide acceptance and integration of GPS for peaceful civil and commercial purposes, promoting GPS as a worldwide standard for international use.[35] "The acceptance of GPS as the world standard . . . enhances the position of the US and allows it to lead in . . . the process of technological and economic globalization."[36] Moreover, "the globalization of GPS markets provides an economic stimulus to firms in the growing US GPS industry, many of which already rely on exports for a significant share of their revenues."[37] Civilian users are not the only economic driver for navigation services as "the potential market for military equipment incorporating satellite navigation is huge."[38] Eventually *all* defense systems will utilize navigation signals, with big money at stake.[39] In terms of the overall defense market, the American defense industry accounts for roughly $100 billion in market capital, with 22 percent representing exports, compared to $50 billion for the EU, with around 25 percent exports.[40] Specifically, "by 2005, the world market [for GPS] is expected to reach $31 billion, 55 percent of which will be outside of the [United States]."[41] Galileo marks the end of the US monopoly on satellite-navigation services; with its planned service upgrades, Galileo could capture a significant share of the market. If Galileo can gain enough market share, it could conceivably threaten, then redefine, the world

IMPLICATIONS AND RECOMMENDATIONS

standard that GPS policy has long sought to control. In order to prevent Europe from introducing an incompatible standard and establishing its own monopoly, the United States argued for international consultations prior to initiating new standards or regulations regarding satellite navigation.[42]

The United States not only fears losing its grasp on the world standard, but also is concerned that the EU (and supporting non-European nations) may pass laws or regulations *mandating* the use of Galileo within certain regions. Since Galileo is partially funded by private investment and is a for-profit enterprise, the EU may be further tempted to compel the use of Galileo by member states or require it for certain purposes in order to generate revenues.[43] EU officials have already publicly asserted that "there will be a transitional period during which Europe will authorize a choice between GPS and Galileo, but EU users eventually will be required to utilize receivers equipped for Galileo," and that "fees and royalties will be levied for use of Galileo chips."[44] An EU mandate could trigger a reactionary US mandate for GPS, needlessly complicating navigation for commercial airlines and other transnational users.

Such pronouncements have spurred the United States to pursue policies which ensure that all satellite-navigation users have the freedom to choose the service (or combination of services) that best meets their needs and to protect the extant GPS-user base.[45] The US position is that GPS users traveling to and from Europe should be required neither to pay Galileo fees nor to install specialized Galileo equipment on boats or airplanes when they can already obtain the same performance from GPS equipment.[46] An EU mandate such as that described above could effectively close the European market to US manufacturers of GPS equipment and cause the United States to take retaliatory measures. For instance, it could respond in kind with mandates or attempt to take legal action against the EU for restraint of fair trade in proper international forums such as the World Trade Organization. Without redress, the current de facto global utility in space navigation would be effectively privatized. And it is not just commercial and civil navigation that would be negatively affected by these potential dictates. A Galileo regional mandate further raises questions regarding NATO and US military equipment in Europe that relies on the GPS M-code. A split, or worse—a declaration requiring Galileo use on European soil—could cause divisions that unravel long-standing US-European military cooperation and integration.

The United States is also concerned that the EU will restrict access to its OS signal specifications. To fully participate in the equipment manufacturing and services markets, US and non-European companies need equal access to technical information.[47] Any restrictions or fees on technical data would either deny American firms access to the Galileo market or unfairly increase their costs. In contrast, the United States openly publishes GPS signal parameters for the SPS (but controls PPS parameters as classified information) at no charge to the public as stipulated in an interface control document (GPS ICD-200), thereby "enabling businesses, scientific and academic institutions, and government entities [worldwide] to develop products, services, and research tools on an equal basis," with no attempt to control the resulting innovations.[48] Like the United States, the EU has no intention of regulating the GNSS applications market, other than carving out a share for Galileo. Whether or not Galileo follows the GPS precedent of openness, the EU will charge royalties "on chipset sales, paid by equipment providers who incorporate a Galileo chip in their products to [access] the Open Service."[49] To earn revenue and control Galileo intellectual-property rights incorporated in the chipsets, the EU may rely on patent protection and may elect to encode the signal, requiring chipsets to contain copyrighted software to decode the OS.[50] A recent EU economic study that analyzed the Galileo business plan deemed the encoding scheme feasible, providing the EU charges all manufacturers the same one-time royalty to facilitate fair competition.[51]

Fair trade is an important concern for the United States. With respect to Galileo, the United States seeks obligations similar to "normal trade relations" (NTR) from Europe.[52] The pressure to produce income for Galileo concessionaires and investors may make it politically difficult for the EU to refrain from imposing tariffs or discriminatory taxes on GPS-related equipment.[53] In particular, the United States disdains any tax on European sales of GPS receivers to provide a revenue source to fund Galileo.[54]

IMPLICATIONS AND RECOMMENDATIONS

Unfair trade practices extend beyond government actions to encompass individual businesses as well, and here the Europeans have some legitimate complaints. At least one American company refused to sell individual cryptological components to Europe and insisted on selling only complete GPS receivers at 10 to 50 times the price.[55]

The advent of Galileo raises additional US concerns regarding technology proliferation. Some of the third-party nations lining up to invest in Galileo are not members of the Missile Technology Control Regime (MTCR), to include China, and there appears to be no safeguard ensuring they will not gain access to advanced space technology that could be applied to missile development and applications. Furthermore, as Galileo shareholders, their privileged access enables them to incorporate satellite-navigation technology into their own domestic weapons programs, including arms manufactured for export. China is already incorporating GPS into its fighter aircraft, and its neighbor and trading partner, North Korea, has reportedly utilized GPS on its submarines.[56] Given access to all phases of satellite-navigation production, launch, control, and operations, these countries could be expected to advance significantly. The North Korean Taepo Dong 2 missile can reach portions of the United States, and the potential addition of satellite-navigation technology would greatly increase its long-range accuracy and amplify the danger of this threat.[57] France sees Galileo as a means to help sell French weapons, further magnifying the potential for technology proliferation.[58] France "cannot sell GPS-supported arms outside of NATO" and chafes at the thought of the United States disapproving sales of French GPS-guided cruise missiles to countries not meeting US criteria.[59] No such requirement could be placed on Galileo-compatible systems. Consequently, controlling technology transfer and proliferation is a primary US goal for cooperation with the EU regarding Galileo.[60]

The US Response

Considering the broad potential ramifications of Galileo's coming operations, the United States has a great deal at stake. Having invested approximately $20 billion in GPS since its inception,

IMPLICATIONS AND RECOMMENDATIONS

not to mention having groomed the concept of satellite navigation to global-utility status, the United States is not about to watch its investment become irrelevant or obsolete.[61] Accordingly, US policy has evolved as Galileo has gained momentum.

Initially, US policy employed a wait-and-see approach towards Galileo, downplaying the need for another system and doubting Europe's ability to pull it off. Officially, the United States saw "no compelling need for Galileo" because GPS would continue to meet the needs of users worldwide.[62] The United States convinced itself that "the availability of GPS without direct charges [would] be enough to win international acceptance of the system" and minimize "the likelihood that competing satellite navigation systems [would] be deployed."[63] It seemed improbable that anyone would pay for a service already available for free. However, as Thomas Schelling explains, "There is a tendency in [US] planning to confuse the unfamiliar with the improbable. The contingency we have not considered looks strange; what looks strange is thought improbable; what is improbable need not be considered seriously."[64] Thus, the United States did not take Galileo seriously, expecting that "plans for a satellite navigation system would be ground to pieces in the gears of the Brussels bureaucracy."[65] As time went by and plans progressed, the United States could not completely ignore Galileo.

In March 1996, the United States reiterated its commitment to providing GPS signals for free and established the Interagency GPS Executive Board.[66] The IGEB, jointly chaired by the Departments of Defense and Transportation, was an effort to downplay the military nature of GPS and strengthen the perception of increased civilian control. In May 2000, the United States stopped degrading GPS civilian accuracy by turning off selective availability in an "effort to make GPS more responsive to civil and commercial users worldwide" and further soften its military image.[67] Then in September 2000, it accelerated GPS modernization by upgrading 12 of the 20 Block IIR satellites to include an additional civilian signal (L2C) and two military signals (M-code).[68] The effort accelerated "the GPS modernization program by approximately eight years" and will eventually raise GPS accuracy on par with Galileo.[69] Ultimately, these efforts failed to increase the international ac-

59

ceptance of GPS and forestall the need for Galileo. In February 1999, the EU announced plans to pursue an independent system, and in March 2002, it obtained approval and funding to launch the Galileo program.[70]

Once the United States accepted that the EU would build Galileo—whether it liked it or not—policy softened from blocking Galileo's progress to ensuring its compatibility and interoperability with GPS.[71] Indicative of this new perspective, the United States announced that it would share its space technology if the EU agreed to a common signal for its OS that would not disrupt the GPS M-code.[72] In particular, the United States recommended a specific signal structure to be shared by Galileo's OS and GPS III. Appealing to European prestige, a member of the US negotiating team called the offer "a major political swing because it says that the US would recognize Galileo as the [international] standard which [GPS III] would follow."[73] In reality, the compromise creates a neutral world standard that Galileo will transmit first since it is scheduled to go operational several years ahead of GPS III. However, the signal that the United States proposes would be slightly less accurate than Galileo's original design.[74] To sweeten the incentive for a common standard, the United States would provide a "favorable view toward export control" on items like space-qualified clocks and radiation-shielded parts, as well as sharing experience in managing large constellations deployed in a very hazardous space environment.[75] In February 2004, Heinz Hilbrecht, EU's chief negotiator for Galileo, responded positively to the US offer, potentially removing the last major obstacle.[76] As the dialogue continues, officials are optimistic that ongoing negotiations will produce a GPS-Galileo cooperation agreement that resolves the technical, trade, and security issues.[77]

Recommendations

"Policy provides the framework within which military and industry leaders can plan for the future."[78] As discussed previously, the current US policy towards Galileo attempts to foster a "cooperative relationship . . . allowing industry to compete in the applications market."[79] However, the "current GPS management, funding and modernization plans are not struc-

IMPLICATIONS AND RECOMMENDATIONS

tured to respond to international competition."[80] Faced with the reality of Galileo, the United States needs to cooperate where it can and compete where it must by continuing efforts to develop a common standard for satellite navigation and taking steps to strengthen the competitiveness of GPS.[81]

A common standard for satellite navigation provides a framework for competition and cooperation, creating "a level playing field for commerce."[82] Jeffrey Bialos, former head of the US delegation for negotiations on the future of GPS and Galileo, likened the utility of satellite navigation to the World Wide Web and argued that it "would make no more sense to have two disconnected, non-interoperable and exclusionary global navigation systems . . . than it would to have two Internets."[83] Standardized competition provides a better product for the user as it leads to more innovative applications and more responsive modernization. Without it, the resulting complacency stagnates development. The lack of a competitor in space "is most assuredly causing complacency in the United States, stunting the expansion of its space capabilities, and further causing [its] allies to develop their own potentially conflicting . . . space capabilities."[84] Arguably without the threat of Galileo, GPS civilian users would still suffer degraded accuracy via selective availability and would not enjoy the benefits of a second civilian signal (L2C) until the launch of Block IIF satellites in mid-2006.[85]

Europe clearly understands that in order to succeed, Galileo must be interoperable and compatible with GPS because "compatibility is the only way to open new applications and to increase market interest in areas in which the existence of two systems offers numerous advantages."[86] Compatibility also implies Galileo will not interfere with or degrade GPS, so development of a common standard will inherently mitigate any signal-overlay issues.[87] From a national security perspective, this is the primary reason the United States should seek compatibility with Galileo.

Beyond healthy competition, a common standard also promotes cooperation. Standardization enables the United States and Europe to exploit the synergy of *combining* GPS and Galileo capabilities. According to a US DOT report, "Using signals from

IMPLICATIONS AND RECOMMENDATIONS

other satellite navigation systems along with GPS . . . offers the potential to enhance integrity, availability, and . . . accuracy for civilian users."[88] The report also deduces that Galileo could effectively mitigate "the consequences of a major GPS system disruption or satellite problem."[89] The EC agreed, noting that "combined performance will provide significant enhancements over individual performances of either GPS or Galileo, opening up applications that would otherwise be impossible for either GPS or Galileo to fulfill alone."[90] Combining the two systems provides users with access to more than 50 satellites, versus 24 or 30 for GPS and Galileo alone, respectively. More satellites mean a higher probability of better satellite geometry—and therefore better accuracy—especially in cities, mountainous and heavily forested terrain, and higher latitudes. Furthermore, a common standard simplifies and streamlines development of combined GPS/Galileo receivers utilizing the same antenna and circuitry.[91]

Continuing the themes of cooperation and compatibility, a 1995 RAND report remarked that "there is no international organization that can address all [GNSS-related] issues."[92] A common standard between GPS and Galileo could provide the foundation for a new international paradigm for global navigation, setting the precedent for potential upgrades to Russia's GLONASS and other emerging systems like China's Beidou constellation.

Besides developing a common standard, the United States must strengthen the competitiveness of GPS if it is to compete successfully and remain viable. A recent *Foreign Affairs* article warned, "Today, through lack of focus and funding, the United States stands to lose not only its primacy but even its capability in satellite navigation if it does not rise to the occasion."[93] Analysis of proposed constellation fill plans indicates that GPS would cede global leadership to Galileo sometime between 2008 and 2010.[94] To remain competitive, the United States must separate the military and civilian aspects of GPS and aggressively pursue upgrades to the latter.

The United States has already taken the first step towards separate military and civilian systems by spectrally separating the M-code from the SPS. Although spectral separation improves the military's ability to deny GPS in localized areas, the separation

IMPLICATIONS AND RECOMMENDATIONS

provides no benefit to civilian users and does little to strengthen GPS competitiveness. However, taking the idea of separation one step further could greatly boost GPS's global viability.

Dividing GPS into autonomous military and civilian systems using the same or slightly modified infrastructures "would enable GPS to address Galileo's challenge more effectively."[95] The Air Force would manage the M-code service, and some civilian organization, possibly the DOT, would manage the SPS. The DOT already maintains a liaison at the GPS Master Control Station.[96] Splitting off the SPS into a separate civilian-managed service would create a civil subset of GPS with the "crucial commercial orientation required to define, develop, and market customer-oriented services."[97] By virtue of its predominant military disposition, GPS competes globally with a decided disadvantage in commercial markets. Freed of the restraints of commercial and civilian requirements, however, the military aspects of GPS could flourish under Air Force leadership, both enhanced by civil advances and complementing them.

Another way in which the United States can strengthen the competitiveness of GPS is to provide comparable services by the time Galileo is expected to begin operation. By accelerating GPS modernization and moving up the launch schedules for the Block IIF and GPS III programs, the United States can reduce Galileo's appeal—but doing so is far beyond current budget outlays for GPS modernization. When Secretary of Defense Donald Rumsfeld inquired about the feasibility of accelerating the GPS III launch schedule by two years (to begin in 2010), he learned it would cost the Air Force an additional $300 million through FY 2009.[98] Separating military and civilian GPS costs could reduce the DOD portion of that increase significantly, while accelerating the development of a competitive civilian GPS infrastructure.

With the DOD (specifically the Air Force) solely in charge of the GPS budget, civilian requirements take a backseat to military priorities—directly encroaching upon the system's global competitiveness. While national security must always come first, the current system creates a zero-sum environment where military and civilian needs compete, usually to the detriment of the latter. The United States created the IGEB in 1996 to increase civilian involvement and establish joint management of GPS between

IMPLICATIONS AND RECOMMENDATIONS

the DOD and DOT. However, the Defense Department remains the main funding and operating agency for GPS and continues to acquire, operate, and maintain all GPS services.[99] Since the US Air Force funds GPS, it must compete with other Air Force space programs and other air platforms for its budget. In 2002, for example, the Air Force contemplated cutting funding and delaying the launch of the GPS III program to bail out the Space-Based Infrared System.[100] In 2003, when the DOD reviewed the possibility of accelerating the GPS III program, the Air Force advised against acquiring GPS improvements ahead of schedule.[101] Maj Gen Franklin J. "Judd" Blaisdell, the Air Force's director of space operations and integration, questioned "whether it is worth the billions and billions of dollars to get it early," and stated that the war fighter appeared satisfied with the state of existing GPS capabilities and planned improvements.[102] The general added, "At this point in time, maybe I have some other things I need to spend my money on."[103] Also in late 2003, Pentagon officials considered delaying procurement and launch of Block IIF satellites to free up $220 million for other Air Force uses in fiscal year 2005.[104] Delays in the procurement of GPS modernization prolong the performance gap between GPS and Galileo, widening Galileo's window of opportunity. As I have shown, the biggest challenge to GPS's competitiveness is the brief window of opportunity for Galileo to capture market share with its enhanced services. It is not advisable to lengthen that window intentionally and increase Galileo's appeal. A separate budget supporting only civilian requirements and managed outside the DOD would help reduce the dilemma of balancing military priorities against civilian needs.[105]

A distinct budget complements the current plan that spectrally separates military and civilian signals as initial steps towards independent military and civilian services and, possibly, separate constellations in the long term. Most space missions have some dual-use aspect, and many have spawned separate satellite constellations to support US military and civilian users. For example, the weather-forecasting mission produced the military Defense Meteorological Satellite Program (DMSP) and the civilian Geostationary Operational Earth Satellite (GOES) system; the two systems work in tandem for the benefit of both groups of users.[106] Separation of satellite communications generated various military systems like the Defense Satellite Communications System (DSCS) and civilian systems like INTELSAT and Inmarsat. Earth observation produced numerous classified military systems as well as civilian programs like France's Satellite Pour L'observation de la Terre (SPOT) and America's Ikonos and Landsat programs. The satellite-navigation mission should not be any different, especially when one considers that the inability of a military system to meet civilian needs in a timely manner served as a major incentive to build Galileo.

The time is undoubtedly not right for the United States to separate military and civilian navigation programs completely. At Euro 3.6 billion, the staggering cost of Galileo is reason enough for the United States to forgo building a civilian sibling to GPS. Alternatively, pooling resources to form a civilian international consortium to manage a system based on GPS SPS and Galileo would distribute the costs, treating satellite navigation like the public good it has become. Separate systems enable military and civilian communities to focus on their primary missions, unhindered by each other's conflicting or non-supportive requirements. Thus, each can concentrate on producing the best systems to meet their unique perspectives.

In conclusion, the proposed Galileo satellite-navigation system challenges US national security and economic productivity. The European system currently impinges on US space superiority because it could interfere with GPS signals and nullify the concept of NAVWAR. Questionable security of the PRS encryption scheme and broad international participation heighten the fear of future hostile use of Galileo against US interests. Economically, Galileo erodes GPS's status as the world standard. The EU's need to generate revenue raises concerns regarding access to signal specifications, fair-trade practices, and proliferation of space technology. In response, the United States must work with the EU to develop a common standard for satellite navigation as a framework for cooperation and competition. Within this framework, the United States must strengthen GPS's competitiveness by (1) accelerating GPS modernization where possible to minimize Galileo's appeal and (2) separating military and civilian services to enable both sectors to minimize conflict within a dual-

IMPLICATIONS AND RECOMMENDATIONS

use system and focus on their specific needs. In this manner, the United States can rise to the occasion, cooperating where it can and competing where it must, to maintain global leadership in satellite navigation and uphold its position in space.

Notes

1. The White House, Office of Science and Technology Policy, "US Global Positioning System."
2. Ibid.
3. Air Force Doctrine Document (AFDD) 1, *Air Force Basic Doctrine*, 85.
4. Hewish, "What Is Happening with GPS?" 57.
5. Richardson, "GPS in the Shadows of Navwar," 23.
6. Ibid.
7. "US Warns EU."
8. Bell, "GPS and Galileo."
9. European Commission, *State of Progress*, 9.
10. Divis, "This Is War," 13.
11. European Commission, *State of Progress*, 13.
12. Karner, interview (see chap. 2, n. 3).
13. de Selding, "Europe Concedes to US," 3.
14. Karner, interview.
15. Ibid.; and Divis, "This Is War," 13.
16. Karner, interview.
17. European Commission, *State of Progress*, 33.
18. Bell, "GPS and Galileo."
19. "USA to Fall behind Europe."
20. European Commission, *State of Progress*, 8.
21. Bell, "GPS and Galileo."
22. European Commission, *State of Progress*, 33.
23. Taverna, "Europe Declares Satnav Independence," 24.
24. European Commission, *State of Progress*, 33–34.
25. Ibid., 34.
26. "Galileo Progress."
27. European Commission, *State of Progress*, 14.
28. "China Joins EU Space Program."
29. Lamoureux, "Opening of EU-China Negotiations."
30. Buck and Dempsey, "China to Join EU's Galileo."
31. "Galileo Progress."
32. Taverna, "Europe Declares Satnav Independence," 24.
33. Taverna and Wall, "Chinese Connection," 23; and Buck and Dempsey, "China to Join EU's Galileo."
34. "China Joins EU Space Program."
35. The White House, Office of Science and Technology Policy, "US Global Positioning System."

36. National Academy of Public Administration, *Global Positioning System*, 42.
37. Ibid.
38. Divis, "Military Role for Galileo Emerges," 12.
39. European Commission, "Galileo: An Imperative for Europe," annex 1, 8.
40. Ibid.
41. National Academy of Public Administration, *Global Positioning System*, 16.
42. Braibanti and Kim, "GPS-Galileo Negotiations," slide 8.
43. "U.S. Officials Cite Concerns."
44. Taverna, "Europe Declares Satnav Independence," 25.
45. Boucher, "State Department Spokesman on GPS."
46. "U.S. Officials Cite Concerns."
47. Boucher, "State Department Spokesman on GPS."
48. Bureau of Public Affairs, "U.S. Global Positioning System."
49. PricewaterhouseCoopers, *Inception Study*, 3.
50. Ibid., 4.
51. PricewaterhouseCoopers, "Galileo Study, Phase II," 18.
52. Braibanti and Kim, "GPS-Galileo Negotiations," slide 13. Also, "NTR is the norm in bilateral trade relationships between countries. Under NTR both parties agree not to extend to any third party nation any trade preferences that are more favorable than those available under the agreement concluded between them unless they simultaneously make the same provisions available to each other" (http://www.itds.treas.gov/mfn.html).
53. Karner, interview. Based on her statement that "the bidders for the concessionaire contract want a guaranteed level of revenue. If they cannot achieve it from the market, they want the EC to make up any difference."
54. Sietzen, "Galileo Takes on GPS," 38–42.
55. European Commission, "Galileo: An Imperative for Europe," annex 1, 8.
56. Snyder, "Navigating the Pacific Rim," chap. 10.
57. Lindstrom and Gasparini, *Galileo Satellite System*, 24.
58. Divis, "Galileo Breaks Free."
59. Ibid.; and Grant and Keohane, "Europe Needs More Space," xiv.
60. Braibanti and Kim, "GPS-Galileo Negotiations," slide 6.
61. "Business: Navigating the Future," 92.
62. Derse, "U.S. Position on Galileo."
63. Lachow, "GPS Dilemma," 141–42.
64. Space Commission, *Report of the Commission*, xv.
65. "USA to Fall behind Europe."
66. Presidential Decision Directive NSTC-6.
67. *Public Papers: Clinton*, 803.
68. Foust, "U.S. Air Force Awarded."
69. Ibid.
70. European Commission, *Galileo: Involving Europe*, iv; and Yoshida, "Complex, Costly Galileo," 1.
71. Singer, "White House Directs Negotiators," 4.

72. Dinmore, "Deal Offered in Satellite," 9.
73. Karner, interview.
74. Knight, "Row over GPS Jamming."
75. Karner, interview.
76. de Selding, "Frequency Concession Removes Galileo," 6.
77. Dempsey, "US and EU Poised," 8.
78. Pace et al., *Global Positioning System*, 204.
79. Karner, interview.
80. Parkinson, "Capability and Management Issues."
81. Ibid.
82. Braibanti and Kim, "GPS-Galileo Negotiations," slide 6.
83. Bialos, "Togetherness on Galileo?" 15.
84. Dolman, *Astropolitik: Classical Geopolitics*," 157.
85. "Boeing to Upgrade GPS."
86. European Parliament, *Report on the Commission Communication to the European Parliament*, 17.
87. Parkinson, "Capability and Management Issues."
88. US Department of Transportation, *Vulnerability of the Transportation Infrastructure*, 49.
89. Ibid.
90. European Commission, "European Dependence," technical note, 22.
91. Braibanti and Kim, "GPS-Galileo Negotiations," slide 11.
92. Pace et al., *Global Positioning System*, 209.
93. Braunschvig et al., "Space Diplomacy," 162.
94. Parkinson, "Capability and Management Issues."
95. Braunschvig et al., "Space Diplomacy," 163.
96. Steven Bayless, program analyst, Office of Navigation Spectrum Policy, Department of Transportation, interview by author, 19 February 2004.
97. Braunschvig et al., 163.
98. Stephens, "Boeing, Lockheed Awarded."
99. Except augmentations to GPS. See Presidential Decision Directive NSTC-6.
100. Divis, "GPS III," 10.
101. Sirak, "Holding the Higher Ground," 21.
102. Ibid.
103. Ibid.
104. Butler, "GPS Spacecraft Lasting Longer," 5.
105. A civilian budget would also need to cover expenses addressing the impact of civil requirements on military requirements.
106. The Air Force transferred control of the DMSP to a joint operational team comprised of the Air Force and the National Oceanic and Atmospheric Administration in May 1998, but they remained separate systems for many years.

Abbreviations

AFDD	Air Force Doctrine Document
AFSCN	Air Force Satellite Control Network
BFT	blue force tracking
bps	bits per second
C/A-code	coarse acquisition code
CS	Commercial Service
DG-TREN	Directorate-General for Energy and Transport
DMSP	Defense Meteorological Satellite Program
DOD	Department of Defense
DOT	Department of Transportation
EC	European Commission
EGNOS	European Geostationary Navigation Overlay Service
ESA	European Space Agency
ESDP	European Security and Defense Policy
EU	European Union
GLONASS	Global Navigation Satellite System
GNSS	global navigation satellite system
GOC	Galileo Operating Company
GOES	Geostationary Operational Earth Satellite
GPS	Global Positioning System
ICBM	intercontinental ballistic missiles
IGEB	Interagency GPS Executive Board
INS	inertial navigation systems
INTELSAT	International Telecommunication Satellite Organization
M-code	military code
MCS	Master Control Station
NATO	North Atlantic Treaty Organization
NAVWAR	navigation warfare
NSCC	navigation system control center
NTR	normal trade relations
OAF	Operation Allied Force

ABBREVIATIONS

OIF	Operation Iraqi Freedom
OS	Open Service
OSS	orbitography and synchronization stations
P-code	precision code
PNT	position, navigation, and timing
PPS	Precise Positioning Service
PRN	pseudorandom noise
PRS	Public Regulated Service
SAR	Search and Rescue
SLBM	submarine launched ballistic missile
SoL	Safety-of-Life
SPS	Standard Positioning Service
TT&C	telemetry, tracking, and commanding
WAAS	Wide Area Augmentation System

Bibliography

Adams, Lt Col Thomas K. "GPS Vulnerabilities." *Military Review* 81, no. 2 (March–April 2001): 10–12.

"Airbus Today." *Airbus*, 26 January 2004. http://www.airbus.com/about/history.asp.

Air Force Doctrine Document (AFDD) 1. *Air Force Basic Doctrine*, 1 September 1997.

Air Force Doctrine Document (AFDD) 2-2. *Space Operations*, 27 November 2001.

Alterman, Dr. Stanley B. "GPS Dependence: A Fragile Vision for US Battlefield Dominance." *Journal of Electronic Defense* 18, no. 9 (September 1995): 52–54.

Bell, Robert G., NATO assistant secretary general for defence support. "GPS and Galileo: Capabilities and Compatibility." Address. European Satellites for Security Conference, Brussels, Belgium, 19 June 2002.

Benedicto, Javier, S. E. Dinwiddy, G. Gatti, R. Lucas, and M. Lugert. "Galileo: Satellite System Design and Technology Developments." *European Space Agency*, November 2000. http://ravel.esrin.esa.it/docs/galileo_world_paper_Dec_2000.pdf (accessed 6 November 2003).

Bialos, Jeffrey P. "Togetherness on Galileo?" *Space News International*, 6 May 2002, 15.

"Boeing to Upgrade GPS 2F Series Birds As Military Needs Increase." *Space Daily*, 19 November 2003. http://www.spacedaily.com/news/gps-03zs.html (accessed 20 February 2004).

Boucher, Richard, US State Department spokesman. "State Department Spokesman on GPS and Galileo." Press briefing, 26 March 2002. United States Mission to the European Union. http://www.useu.be/Galileo/Mar2602BoucherGalileoGPS.html (accessed 5 November 2003).

Braibanti, Ralph, and Jason Y. Kim. Briefing to US GPS Industry Council, Sunnyvale, CA, 21 March 2002. Subject: GPS-Galileo Negotiations: Commercial Issues at Stake. http://www.technology.gov/space/library/speeches/2002-04-24-ISAC-briefing.ppt.

Braunschvig, David, Richard L. Garwin, and Jeremy C. Marwell. "Space Diplomacy." *Foreign Affairs*, July/August 2003, 159–62.

BIBLIOGRAPHY

Buck, Tobias, and Judy Dempsey. "China to Join EU's Galileo Navigation Plan." *London Financial Times*, 19 September 2003.

Bulmahn, Edelgard, federal minister of education and research (Germany). "Europe's Ambitions in Space." Address. Center for International Science and Technology Policy, George Washington University, Washington, DC, 6 February 2002.

Bureau of Public Affairs, US Department of State. "U.S. Global Positioning System and European Galileo System." Media note. Washington, DC: Office of the Spokesman, 7 March 2002. http://www.state.gov/r/pa/prs/ps/2002/8673.htm (accessed 5 November 2003).

"Business: Navigating the Future; GPS and Galileo." *Economist* 367, no. 8330 (28 June 2003): 92.

Butler, Amy. "GPS Spacecraft Lasting Longer Than Expected, Prompting Possible IIF Delay," *Defense Daily*, 2 December 2003, 5.

———. "Rice Wants President to Initiate Sweeping Space Policy Review." *InsideDefense.com*, 14 May 2002. http://www.californiaspaceauthority.org/pr020517a.html (accessed 16 January 2004).

Byrne, Gerry. "Global Fix: What's So Wrong with GPS That Europe's Spending Billions on an Alternative? Plenty." *New Scientist* 174, no. 2341 (May 2002): 32–36.

"China Joins EU Space Program to Break US GPS Monopoly." *Space Daily*, 27 September 2003. http://www.spacedaily.com/news/gps-03zc.html.

Clinton, William J. "Statement on the Decision to Stop Degrading Global Positioning System Signals." *Public Papers of the Presidents of the United States: William J. Clinton, 2000–2001*. Book 1, 1 May 2000. Washington, DC: Government Printing Office, 2001. http://www.presidency.ucsb.edu/ws/index.php?pid=58423&st=&st1=.

Commission of the European Communities. *Commission Communication to the European Parliament and the Council: On Galileo*. COM(2000) 750. Brussels: European Commission, 22 November 2000. http://europa.eu.int/comm/dgs/energy_transport/galileo/doc/gal_com_2000_750_en.pdf (accessed 31 October 2003).

Dempsey, Judy. "US and EU Poised to Agree on Satellite Navigation Networks." *London Financial Times*, 3 February 2004, 8.

de Palacio, Loyola, vice president of the European Commission. "The Importance of Galileo for Europe." Address. Internationaler Kongress Kommerzielle Anwendung der Satelliten-Navigation. Munich, Germany, 26 April 2001. http://eu.spaceref.com/news/viewpr.html?pid=5419 (accessed 5 December 2004).

Derse, Anne, minister counselor for economic affairs at the US Mission to the European Union. "U.S. Position on Galileo." Statement to *Euronews*, 17 January 2002. http://www.useu.be/Galileo/Jan2902USGalileo.html (accessed 5 November 2003).

de Selding, Peter B. "Europe Concedes to US on One GPS, Galileo Roadblock." *Space News*, 18 December 2003, 3.

———. "Frequency Concession Removes Galileo Agreement Roadblock." *Space News*, 9 February 2004, 6.

Dinerman, Taylor. "GPS and Galileo: European Illusions and American Fears." *Ad Astra* 14, no. 2 (March/April 2002): 10.

Dinmore, Guy. "Deal Offered in Satellite Navigation Row." *London Financial Times*, 9 January 2004, 9.

Divis, Dee Ann. "Galileo Breaks Free." *GPS World*, 1 July 2003. http://www.gpsworld.com/gpsworld/article/articleDetail.jsp?id=63243 (accessed 28 December 2003).

———. "GPS III, Modernization Face Budget Cut." *GPS World* 13, no. 7 (July 2002): 10–11.

———. "Military Role for Galileo Emerges." *GPS World* 13, no. 5 (May 2002): 10, 12–13.

———. "This Is War: FCC's Disputed Limits." *GPS World*, 1 May 2003. http://www.gpsworld.com/gpsworld/articleDetail.jsp?id=56991 (accessed 28 December 2003).

Dolman, Everett C. *Astropolitik: Classical Geopolitics in the Space Age.* London: Frank Cass Publishers, 2002.

Dornheim, Michael A. "GPS Improvements Set to Help Civil Users." *Aviation Week and Space Technology* 157, no. 13 (23 September 2002): 56.

Dunn, Richard J., III. *Blue Force Tracking: The Afghanistan and Iraq Experience and Its Implications for the US Army.* Reston, VA: Northrop Grumman Mission Systems, 2003.

BIBLIOGRAPHY

http://www.analysiscenter.northropgrumman.com/files/BFT-WP%20Halfc.pdf (accessed 9 December 2005).

"EU Law + Policy Overview." *European Union in the US.* http://www.eurunion.org/legislat/Defense/esdpweb.htm (accessed 20 February 2004).

European Commission. Communication from the Commission to the European Parliament and the Council. *State of Progress of the Galileo Programme.* COM (2002) 518 final, 24 September 2002. http://europa.eu.int/comm/dgs/energy_transport/galileo/doc/com2002_518_en.pdf (accessed 1 April 2004).

———. Directorate-General for Energy and Transport. "Galileo: A Decision Must Be Taken Urgently." Information note, March 2002.

———. Directorate-General for Energy and Transport. "Galileo: An Imperative for Europe." Information note. Annex 1, 18 January 2002. http://www.tvlink.org/pdf/galileo_en_final.pdf.

———. *The European Community: Crossroads in Space.* EC Report EUR 14010. Luxembourg: Office for Official Publications of the EC, 1991.

———. *The European Dependence on US-GPS and the Galileo Initiative.* Brussels, Belgium: Directorate-General for Energy and Transport, 8 February 2002. http://europa.eu.int/comm/dgs/energy_transport/galileo/doc/gal_european_dependence_on_gps_rev22.pdf (accessed 31 October 2003).

———. "The European Dependence on US-GPS and the Galileo Initiative." Technical note. Brussels, Belgium: European Commission, Directorate-General for Energy and Transport, 14 February 2002.

———. *The European Union and Space: Fostering Applications, Markets and Industrial Competitiveness.* EC Directorate-General XII: Science, Research and Development. Luxembourg: Office for Official Publications of the EC, 1997.

———. *Galileo: Involving Europe in a New Generation of Satellite Navigation Services.* COM (1999) 54 Final. Brussels, Belgium: European Commission, 10 February 1999. http://europa.eu.int/comm/dgs/energy_transport/galileo/doc/com_1999_54_en.pdf (accessed 31 October 2003).

──────. *Galileo: Mission High Level Definition* (Mission Requirements Document). Version 3.0, 23 September 2002. http://europa.eu.int/comm/dgs/energy_transport/galileo/doc/galileo_hld_v3_23_09_02.pdf (accessed 31 October 2003).

──────. *The Galileo Project: Galileo Design Consolidation*. August 2003. http://europa.eu.int/comm/dgs/energy_transport/galileo/doc/galilei_brochure.pdf (accessed 31 October 2003).

──────. "Towards a Coherent European Approach for Space." EC Working Document, SEC(1999) 789, 7 June 1999.

"Europe: Eppur Si Muove—or Maybe Not; Europe's Galileo Satellite Positioning System." *Economist* 363, no. 8275 (1 June 2002): 53–54.

European Parliament. *Report on the Commission Communication to the Council and the European Parliament on Europe and Space: Turning to a New Chapter*. Final A5-0451/2001. Committee on Industry, External Trade, Research and Energy, 6 December 2001. http://www.europarl.eu.int/registre/seance_pleniere/textes_deposes/rapports/2001/0451/P5_A(2001)0451_EN.doc.

──────. *Report on the Commission Communication to the European Parliament and the Council: On Galileo*. EP Report A5-0288/2001. Brussels, Belgium: European Parliament Committee on Regional Policy, Transport and Tourism, 20 July 2001.

European Space Agency. "Why Europe Needs Galileo." http://www.esa.int/export/esaSA/GGG0H750NDC_navigation_0.html (accessed 6 November 2003).

Federation of American Scientists. "Military Space Programs: Transit," 13 April 1997. http://www.fas.org/spp/military/program/nav/transit.htm (accessed 29 December 2003).

Foust, Jeff. "The U.S. Air Force Awarded a $53 Million Contract to Lockheed Martin Monday to Begin Upgrades on Global Positioning System." *Space.com*, 12 September 2000. http://209.73.219.100/businesstechnology/technology/gps_modernized_000912.html (accessed 4 November 2003).

Frutkin, Arnold W. *International Cooperation in Space*. Englewood Cliffs, NJ: Prentice-Hall, 1965.

"Galileo Progress: New Alliances, ITTs." *GPS World*, 1 November 2003. http://www.gpsworld.com/gpsworld/article/articleDetail.jsp?id=75540 (accessed 28 December 2003).

General Nathan Bedford Forrest Historical Society. "Quotes by General Forrest," n.d. http://www.tennessee-scv.org/ForrestHistSociety/quotes.html (accessed 3 February 2004).

Gibbons, Glen. "Birth of Galileo: A World of Difference." *GPS World*, 2 February 2002. http://www.gpsworld.com/gpsworld/article/articleDetail.jsp?id=118824.

GPS Joint Program Office. "Navstar GPS Fact Sheet," March 2003. http://gps.losangeles.af.mil/jpo/gpsoverview.htm (accessed 14 January 2004).

Grant, Charles, and Daniel Keohane. "Europe Needs More Space." *Britain in Space. New Statesman* 15, no. 707 (Special Supplement) (20 May 2002): xiv.

Halsall, Paul, ed. "The Crime of Galileo: Indictment and Abjuration of 1633." *Internet Modern History Sourcebook*, January 1999. http://www.fordham.edu/halsall/mod/1630galileo.html (accessed 25 February 2004).

Handberg, Roger B., and Joan Johnson-Freese. *The Prestige Trap*. Dubuque, IA: Kendal/Hunt Publishing, 1994.

Hewish, Mark. "What Is Happening with GPS?" *Jane's International Defense Review* 36 (July 2003): 53–58.

Knight, Will. "Row over GPS Jamming Still Divides US and Europe." *NewScientist.com*, 3 February 2004. http://www.newscientist.com/news/print.jsp?id=ns99994641.

Kupchan, Charles. *The End of the American Era*. New York: Alfred A. Knopf, 2002.

Lachow, Irving. "The GPS Dilemma." *International Security* 20, no. 1 (Summer 1995): 126–48.

Lamoureux, Francois, director general, European Commission, Directorate-General for Energy and Transport (EC DG-TREN). "Opening of EU-China Negotiations on Satellite Navigation." Address. Brussels, Belgium, 16 May 2003.

Lembke, Johan. *Competition for Technological Leadership: EU Policy for High Technology*. Cheltenham, UK: Edward Elgar Publishing, 2002.

———. "EU Critical Infrastructure and Security Policy: Capabilities, Strategies and Vulnerabilities." *Current Politics and Economics of Europe* 11, no. 2 (2002): 99–129.

———. "The Politics of Galileo." European Policy Paper no. 7. Pittsburgh, PA: University of Pittsburgh, University Center for International Studies, April 2001.

Lindstrom, Gustav, and Giovanni Gasparini. *The Galileo Satellite System and Its Security Implications.* Occasional Paper no. 44. Paris, France: EU Institute for Security Studies, April 2003.

Logsdon, Tom. *The Navstar Global Positioning System.* New York: Van Nostrand Reinhold, 1992.

MacDonald, Keith. "Econosats: Toward an Affordable Global Navigation Satellite System." *GPS World*, September 1993, 44–54.

Mackenzie, Richard. "Apache Attack." *Air Force Magazine Online*, October 1991. http://www.afa.org/magazine/perspectives/desert_storm/1091apache.html (accessed 31 January 2004).

McDougall, Walter A. *. . . the Heavens and the Earth: A Political History of the Space Age.* Baltimore, MD: Johns Hopkins University Press, 1985.

McLucas, John L. *Space Commerce.* Cambridge, MA: Harvard University Press, 1991.

National Academy of Public Administration and the National Research Council. *The Global Positioning System: Charting the Future.* Report for the US Congress and Department of Defense. Washington, DC: National Academy Press, 1995.

National Research Council. *The Global Positioning System: A Shared National Asset.* Washington, DC: National Academy Press, 1995.

"Navstar Global Positioning System." *Federation of American Scientists.* http://www.fas.org/spp/military/program/nav/gps.htm (accessed 3 November 2003).

Pace, Scott, Gerald Frost, Dave Freligner, Donna Fossum, Donald K. Wassem, and Monica Pinto. *The Global Positioning System: Assessing National Policies.* Santa Monica, CA: RAND, 1995.

Pappas, Maj Zannis M. "Effects of the Galileo Constellation on US National Interests." Research Report no. 32-1229. Maxwell AFB, AL: Air Command and Staff College, April 2003.

Parkinson, Dr. Brad. "Capability and Management Issues for GPS/Galileo Positioning and Timing." Presentation. Council on Foreign Relations, 7 November 2003.

Peartree, C. Edwards, C. Kenneth Allard, and Carl O'Berry. "Information Superiority." In *Air and Space Power in the New Millennium*. Edited by Daniel Gouré and Christopher M. Szara. Washington, DC: Center for Strategic and International Studies, 1997.

Presidential Decision Directive NSTC-6. The White House. To the vice president et al. Memorandum, 28 March 1996. http://www.schriever.af.mil/GpsSupportCenter/documents/gps_pdd.htm (accessed 14 January 2004).

PricewaterhouseCoopers. "Galileo Study, Phase II Executive Summary." Belgium: Corporate Finance Division, 17 January 2003.

———. *Inception Study to Support the Development of a Business Plan for the GALILEO Programme*. TREN/B5/23-2001. Executive Summary. Belgium: Corporate Finance, 20 November 2001. http://europa.eu.int/comm/dgs/energy_transport/galileo/doc/gal_exec_summ_final_report_v1_7.pdf (accessed 31 October 2003).

Reid, T. R. "EU Summit Ends with a Bang and a Whimper." *Washington Post*, 17 March 2002.

Richardson, Doug. "GPS in the Shadows of Navwar." *Armada International* 22, no. 4 (August–September 1998): 23–26.

Rip, Michael R., and James M. Hasik. *The Precision Revolution: GPS and the Future of Aerial Warfare*. Annapolis, MD: Naval Institute Press, 2002.

Robertson, George Islay MacNeill (Lord Robertson). "NATO in the 21st Century." Speech. Secretary General to the Millennium Year Lord Mayor's Lecture, 20 July 2000. http://www.nato.int/docu/speech/2000/s000720a.htm (accessed 30 January 2004).

Robinson, Bruce T. "Who Goes There?" *IEEE Spectrum Online*, 1 October 2003. http://www.spectrum.ieee.org/WEBONLY/publicfeature/oct03/mili.html (accessed 19 January 2004).

Roos, John G. "A Pair of Achilles' Heels." *Armed Forces Journal International*, November 1994, 21–23.

Sebesta, Lorenza. "US-European Relations and the Decision to Build Ariane, the European Launch Vehicle." In *Beyond the Ionosphere: Fifty Years of Satellite Communication.* Edited by Andrew Buttrica. NASA SP-4217. Washington, DC: NASA, 1997.

Sietzen, Frank, Jr. "Galileo Takes on GPS." *Aerospace America* 41, no. 8 (August 2003): 38–42.

Singer, Jeremy. "White House Directs Negotiators to Ease Stance on Galileo Program." *Space News*, 26 August 2002, 4.

Sirak, Michael. "Holding the Higher Ground." *Jane's Defence Weekly* 40, no. 14 (8 October 2003): 21.

———. "USA Sets Sights on GPS Security Enhancements." *Jane's Defence Weekly*, 16 January 2002, 30.

Snyder, Amy Paige. "Navigating the Pacific Rim: Regional Satellite Navigation/Positioning Capabilities and Relevant Policy Issues for the United States." In *Space and Military Power in East Asia: The Challenge and Opportunity of Dual-Purpose Space Technologies*. Edited by Rebecca Jimerson and Ray A. Williamson. Washington, DC: Space Policy Institute, December 2001. http://www.gwu.edu/~spi/gpspaper.html.

Space Commission. *Report of the Commission to Assess United States National Security Space Management and Organization.* Washington, DC: Space Commission, 11 January 2001.

Stephens, Hampton. "Boeing, Lockheed Awarded $20 Million Each for GPS Study Phase." *Inside the Air Force*, 9 January 2004.

Taverna, Michael A. "European Challenger Go-Ahead for Galileo May Force U.S. Leaders to Stake a New Position on Satnav System." *Aviation Week and Space Technology* 159, no. 10 (8 September 2003): 61.

———. "Europe Declares Satnav Independence: Europe and the US Must Now Discuss Thorny Political and Technical Issues Raised by Galileo Approval." *Aviation Week and Space Technology* 156, no. 13 (1 April 2002): 24.

Taverna, Michael A., and Robert Wall. "The Chinese Connection: Beijing's Plan to Join Europe's Galileo Satnav System Could Draw a Rebuff from the U.S." *Aviation Week and Space Technology* 159, no. 13 (29 September 2003): 23, 26.

United States Mission to the European Union. "US Global Positioning System Policy Fact Sheet," 28 March 1996. http://www.useu.be/Galileo/Mar281996GPSFactSheet.html.

"USA to Fall behind Europe in Space When Galileo Operational." *der Spiegel* Web site, 27 October 2003. Provided by BBC Worldwide Monitoring.

US Department of Transportation. *Vulnerability of the Transportation Infrastructure Relying on the Global Positioning System* (Volpe Report). Cambridge, MA: John A. Volpe National Transportation Systems Center, 29 August 2001.

"U.S. Officials Cite Concerns about Planned European Satellite System." Brussels, Belgium: United States Mission to the European Union, 12 February 2002. http://www.useu.be/Galileo/Feb1202GalileoBraibanti.html (accessed 5 November 2003).

"US Warns EU about Galileo's Possible Military Conflicts." *Space Daily*, 18 December 2001. http://www.spacedaily.com/news/gps-euro-01g.html (accessed 19 November 2003).

Warner, Randy. "GPS ICD-200: What Is It and Where Can I Find It?" 8 June 2001. http://www.synergy-gps.com/GPS_ICD-200_Info.PDF (accessed 29 December 2003).

Wertz, James R., and Wiley J. Larson. *Space Mission Analysis and Design*. AA Dordrecht, Netherlands: Kluwer Academic Publishers, 1991.

The White House. Office of Science and Technology Policy. National Security Council. "US Global Positioning System Policy Fact Sheet," 29 March 1996. http://www.ostp.gov/NSTC/html/pdd6.html (accessed 7 November 2003).

Wilson, Andrew, ed. *Galileo: The European Programme for Global Navigation Services*. AG Noordwijk, Netherlands: ESA Publications Division, May 2002.

Yoshida, Junko. "Complex, Costly Galileo Has Implications for U.S. Industry, Military—Europe Looks to One-Up GPS." *Electronic Engineering Times*, 1 April 2002, 1.

GPS versus Galileo
Balancing for Position in Space

Air University Press Team

Chief Editor
Jeanne K. Shamburger

Copy Editor
Tammi K. Long

Book Design and Cover Art
Daniel M. Armstrong

*Composition and
Prepress Production*
Mary P. Ferguson

Print Preparation
Joan Hickey

Distribution
Diane Clark

www.ingramcontent.com/pod-product-compliance
Lightning Source LLC
Chambersburg PA
CBHW081240180526
45171CB00005B/483